Artificial Intelligence in Label-free Microscopy

Ata Mahjoubfar • Claire Lifan Chen • Bahram Jalali

Artificial Intelligence in Label-free Microscopy

Biological Cell Classification by Time Stretch

Springer

Ata Mahjoubfar
University of California Los Angeles
Los Angeles, CA, USA

Claire Lifan Chen
University of California Los Angeles
Los Angeles, CA, USA

Bahram Jalali
University of California Los Angeles
Los Angeles, CA, USA

ISBN 978-3-319-51447-5 ISBN 978-3-319-51448-2 (eBook)
DOI 10.1007/978-3-319-51448-2

Library of Congress Control Number: 2017937310

Printed on acid-free paper

This Springer imprint is published by Springer Nature
The registered company is Springer International Publishing AG
The registered company address is: Gewerbestrasse 11, 6330 Cham, Switzerland

Preface

High-throughput real-time imaging and vision systems for capture and identification of fast phenomena are among the most essential tools for scientific, industrial, military, and most importantly biomedical applications. The key challenge in these instruments is the fundamental trade-off between speed and sensitivity of the measurement system due to the limited signal energy collected in each measurement window. Based on two enabling technologies, namely, photonic time stretch and optical amplification, several novel high-throughput optical measurement tools are recently developed for applications such as volumetric scanning, vibrometry, and flow cytometry. Here, we introduce time stretch imaging, a high-content computer vision system developed for big data acquisition and analysis of images. Time stretch imaging is able to capture quantitative optical phase and intensity images simultaneously, enabling accurate surface inspection, volumetric scans, defect detection, cell analysis, and cancer diagnostics.

We further describe a complete artificial intelligence pipeline for time stretch microscopy that performs optical phase measurement, image processing, feature extraction, and classification. Multiple biophysical features such as morphological parameters, optical loss characteristics, and protein concentration are measured on individual biological cells. These biophysical measurements form a hyperdimensional feature space in which supervised learning is performed for cell classification. The technology is in clinical testing for blood screening and circulating tumor cell detection, as well as studying lipid-accumulating algal strains for biofuel production. By integrating machine learning with high-throughput quantitative imaging, this system achieves record-high accuracy in label-free cellular phenotypic screening and opens up a new path to data-driven diagnosis.

Furthermore, we explained a real-time image compression technique performed in the optical domain to solve the big data challenge created by ultrafast measurement systems. Many ultrafast and high-throughput data acquisition equipment, including time stretch imaging, produce a torrent of data in a short time, e.g., several gigabytes per second. Such a data volume and velocity place a burden on data acquisition, storage, and processing and call for technologies that compress images in optical domain and in real-time. As a solution, we have experimentally demonstrated warped time stretch, which offers variable spectral-domain sampling rate, as well as the ability to engineer the time-bandwidth product of the signal's envelope to match that of the data acquisition systems. We also show how to design the kernel of the transform and, specifically, the nonlinear group delay profile governed by the signal sparsity. Such a kernel leads to smart detection with nonuniform spectral resolution, having direct utility in improvement of data acquisition rate, real-time data compression, and enhancement of ultrafast data capture accuracy.

Los Angeles, CA, USA

Ata Mahjoubfar
Claire Lifan Chen
Bahram Jalali

Acknowledgements

We would like to thank all of our publication coauthors, which our collaborative works consist many chapters of this book. Chapter 2 is a version of IEEE 49th Annual Conference on Information Sciences and Systems (CISS) Proceedings, pp. 7086896, 2015 [1]. Chapter 3 is a version of Applied Physics Letters, Vol. 98, No. 10, 101107, 2011 [12]. Chapter 4 is a version of Proceedings of SPIE, Frontiers in Ultrafast Optics: Biomedical, Scientific, and Industrial Applications XIII, San Francisco, CA, 2013, 86110N [16]. Chapter 5 is a version of Biomedical Optics Express, Vol. 4, No. 9, pp. 1618–1625, 2013 [17]. Chapters 6 and 8 are based on Scientific Reports, Vol. 6, pp. 21471, 2016 [3]. Chapter 7 is a version of Proceedings of SPIE, High-Speed Biomedical Imaging and Spectroscopy: Toward Big Data Instrumentation and Management II, San Francisco, CA, 2017, 100760J [2]. Chapter 9 is a version of PLoS ONE, Vol. 10, No. 4, pp. e0125106, 2015 [104]. Finally, Chapter 10 is a version of Scientific Reports, Vol. 5, pp. 17148, 2015 [4].

Contents

List of Figures

List of Tables

Part I
Time Stretch Imaging

Chapter 1
Introduction

High-throughput measurement instruments are indispensable tools to acquire large volume of data for detection and classification of rare events. Enabled by the photonic time stretch, a new class of instruments with record throughputs have led to the discovery of optical rogue waves, detection of rare cancer cells with record accuracy, and the highest analog-to-digital conversion performance ever achieved.

Here, we work toward introducing time stretch quantitative phase imaging (TS-QPI) [3], a high-throughput label-free imaging flow cytometer developed for analyzing large population of cells. Label-free cell analysis is essential to personalized genomics, cancer diagnostics, and drug development as it avoids adverse effects of staining reagents on cellular viability and cell signaling. However, currently available label-free cell assays mostly rely only on a single feature and lack sufficient differentiation. Also, the sample size analyzed by these assays is limited due to their low throughput. However, TS-QPI is able to capture quantitative optical phase and intensity images simultaneously, enabling high-content cell analysis and phenotypic screening.

We further developed a complete machine learning pipeline that performs optical phase measurement, image processing, feature extraction, and classification. Multiple biophysical features such as morphological parameters, optical loss characteristics, and protein concentration are measured on individual cells. These biophysical measurements form a hyperdimensional feature space in which supervised learning is performed for cell classification. The technology is in clinical testing for blood screening, as well as lipid accumulating algal strains for biofuel production. By integrating machine learning with high-throughput quantitative imaging, this system achieves record high accuracy in label-free cellular phenotypic screening and opens up a new path to data-driven diagnosis.

Furthermore, the instrument's ultrahigh throughput creates a big data challenge. TS-QPI, as well as many other ultrafast measurement system, produces a large volume of data in a short time in the order of tens of Terabytes per hour. Such a data firehose places a burden on data acquisition, storage, and processing and

A. Mahjoubfar et al., *Artificial Intelligence in Label-free Microscopy*,
DOI 10.1007/978-3-319-51448-2_1

calls for technologies that compress images in optical domain and in real-time. As a solution, we have demonstrated, for the first time, real-time image compression performed in the optical domain. Called as warped time stretch [4–6], this technique offers variable rate spectral domain sampling, as well as the ability to engineer the time-bandwidth product of the signal's envelope to match that of the data acquisition systems. We also show how to design the kernel of the transform and specifically, the nonlinear group delay profile governed by the signal sparsity. Such a kernel leads to smart detection with nonuniform spectral resolution, having direct utility in improvement of data acquisition rate, real-time data compression, and enhancement of ultrafast data capture accuracy.

In Chap. 2, an introductory background on photonic time stretch imaging, dispersive Fourier transform, and label-free cell classification is given.

Chapter 3 shows how time stretch imaging can be employed in nanometer-scale optical path length measurements. As a demonstration of the applications, acoustic vibrations on the surface of a Silicon wafer are captured. Chapter 4 extends this capability to laser scanners and enables them to achieve ultrafast volumetric three-dimensional inspection.

In Chap. 5, we demonstrate a high-throughput label-free cellular imaging system using time stretch quantitative phase imaging (TS-QPI). Label-free cell analysis is essential to personalized genomics, cancer diagnostics, and drug development as it avoids adverse effects of staining reagents on cellular viability and cell signaling. However, currently available label-free cell assays mostly rely only on a single feature and lack sufficient differentiation. Also, the sample size analyzed by these assays is limited due to their low throughput. Here, with TS-QPI, quantitative optical phase and loss images of suspended cells are captured simultaneously at flow speeds as high as a few meters per second. This system enables analyzing large population of cells with spatial information, filling the technological gap between flow cytometry and microscopy.

In Chap. 6, a data processing and image analysis pipeline has been developed for time stretch quantitative phase imaging. It performs signal processing, optical phase demodulation, image recognition, and feature extraction on large amount of cell images. Multiple biophysical attributes such as morphological features, optical loss characteristics, and protein concentration are measured on individual cells. The pipeline combining with TS-QPI is in clinical testing for blood screening and has also been applied to classification of high lipid-content algal cell strains for biofuel development.

Chapter 7 outlines and analyzes a data acquisition system for time stretch imaging that enables big data recording at rates of few gigabytes of images per second.

We further expanded the data analytics into a complete machine learning pipeline in Chap. 8. In addition to optical phase measurement, image processing, and feature extraction, the multivariate biophysical features form a hyperdimensional feature space in which supervised learning is performed for cell classification. By integrating artificial intelligence with high-throughput quantitative imaging, we achieve record high accuracy in label-free cellular phenotypic screening and open a new path to data-driven diagnosis.

Furthermore, as a solution to the challenges in storage and analysis created in the big data volume acquired by the ultrahigh throughput instruments, we demonstrated a real-time optical image compression technique. In Chap. 9, the introduced method exploits the sparsity of the image generated with time stretch microscopy, and then reduces the number of samples and the amount of data in the experiments by several times. This data compression performed in the optical domain, also known as warped time stretch or warped stretch transform, functions as a nonuniform sampling technique before analog to digital conversion in time stretch camera, addressing the big data predicament in such systems.

In Chap. 10, we further explore optical data compression technique and show how to design the kernel of the warped stretch transform and specifically, the nonlinear group delay profile governed by the signal sparsity. With the designed kernels, the warped stretch transform offers variable rate spectral domain sampling, as well as the ability to engineer the time-bandwidth product of the signal's envelope to match that of the data acquisition systems. Such a kernel leads to smart detection with nonuniform spectral resolution, having direct utility in improvement of data acquisition rate, real-time data compression, and enhancement of ultrafast data capture accuracy.

Chapter 2
Time Stretch

Time stretch is the leading technology in ultrafast big-data acquisition. Here we introduce time stretch technique and highlight its applications in the context of imaging.

2.1 Time Stretch Imaging

High-throughput optical sensing are indispensable tools to acquire large data sets for detection and classification of rare events. As noninvasive instrument, high-speed optical sensing are widely used in scientific, industrial, military, and biomedical applications. However, as the acquisition speed increases, the signal energy collected in single measurement drops. This leads to a reduction in the signal-to-noise ratio of the measurement, which ultimately limits the resolution and sensitivity of the sensing or imaging application. For optical instrument like camera, one way to collect more photons in each measurement is to increase the intensity of the illumination or the interrogation light, but this is often undesirable in biological applications because the biological samples can easily get damaged by the intense light, especially when an objective lens is focusing the light on the specimen.

Enabled by the photonic time stretch and optical amplification, a new class of instruments with record throughputs have led to the discovery of optical rogue waves, detection of rare cancer cells with record accuracy, and the highest analog-to-digital conversion performance ever achieved. One example of these instruments is the time stretch microscopy, an imaging and sensing modality that features continuous operation at about 100 million frames per second and shutter speed of less than a nanosecond.

Telecommunication systems routinely generate, capture, and analyze data at rates exceeding billions of bits per second. Interestingly, the scale of the problem is similar to that of blood analysis. With approximately 1 billion cells per milliliter

A. Mahjoubfar et al., *Artificial Intelligence in Label-free Microscopy*,
DOI 10.1007/978-3-319-51448-2_2

of blood, detection of a few abnormal cells in a blood sample translates into a "cell error rate" of 10^{-12}, a value that is curiously similar to the bit error rate in telecommunication systems. This suggests that data multiplexing, capture, and processing techniques developed for data communication can be leveraged for biological cell classification.

Time stretch dispersive Fourier transform is a method for real-time capture of ultra wideband signals. It allows acquisition of single shot optical spectra continuously and at tens to hundreds of million frames per second. It has led to the discovery of optical Rogue waves [7] and, when combined with electro-optic conversion, to record analog-to-digital conversion performance [8]. Combination of the telecommunication technique of wavelength division multiplexing (WDM) and the time stretch technique [9], the time stretch camera known as STEAM [10–16] has demonstrated imaging of cells with record shutter speed and continuous throughput leading to detection of rare breast cancer cells in blood with one-in-a-million sensitivity [16–20]. A second data communication inspired technique called fluorescence imaging using radio frequency-tagged excitation (FIRE) is a new approach to fluorescent imaging that is based on wireless communication techniques [21]. FIRE has achieved real-time pixel readout rates one order of magnitude faster than the current gold standard in high-speed fluorescence imaging [21]. Producing data rates as high as one tera bit per second, these real-time instruments pose a big data challenge that overwhelms even the most advanced computers [22]. Driven by the necessity of solving this problem, we have recently introduced and demonstrated a categorically new data compression technology [23, 24]. The so-called Anamorphic (warped) Stretch Transform is a physics based data processing technique that aims to mitigate the big data problem in real-time instruments, in digital imaging, and beyond [22–26]. This compression method is an entirely different approach to achieving similar functionalities as compressive sensing [27, 28] and is more amenable to fast real-time operation.

The operating principle of time stretch imaging is shown in Fig. 2.1. First, the object image is encoded in the spectrum of ultrafast optical pulses. Then, pulses are stretched in time by dispersive Fourier transformation, so that different wavelength components reach a single-pixel photodetector at different times. The time stretch function allows ultrafast image frames to be digitized in real-time. Images are optically amplified before detection and digitization to overcome the thermal noise. The basic principle of time stretch imaging (STEAM) involves two steps both performed optically. In the first step, the spectrum of a broadband optical pulse is converted by a spatial disperser into a rainbow that illuminates the target. Therefore, the spatial information (image) of the object is encoded into the spectrum of the resultant reflected or transmitted rainbow pulse. A one-dimensional or two-dimensional rainbow is used to acquire a line-scan. The 2D image is obtained by scanning the one-dimensional rainbow in the second dimension or by a two-dimensional rainbow. In the second step, the spectrum of the image-encoded pulse is mapped into a serial temporal signal that is stretched in time to slow it down such that it can be digitized in real-time [9]. This optically amplified time stretched serial stream is detected by a single-pixel photodetector and the

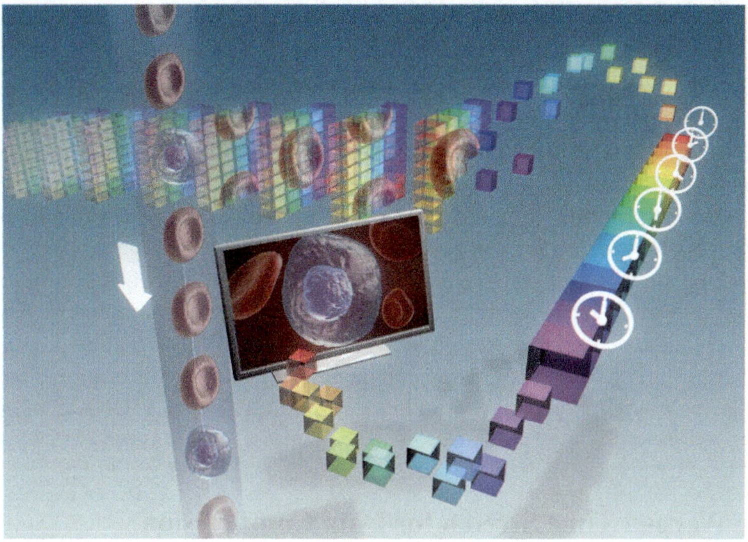

Fig. 2.1 Operating principle of time stretch imaging. The key innovations in STEAM that enable high-speed real-time imaging are photonic time stretch for digitizing fast images in real-time and the optical image amplification for compensating the low number of photons collected during the ultra-short shutter time

image is reconstructed in the digital domain. Subsequent pulses capture repetitive frames. The laser pulse repetition rate corresponds to the frame rate and the temporal width of the pulses corresponds to camera's shutter speed (exposure time). The key innovations in STEAM that enable high-speed real-time imaging are photonic time stretch for digitizing fast images in real-time and the optical image amplification for compensating the low number of photons collected during the ultra-short shutter time [29].

2.2 Cell Classification Using Time Stretch Imaging

Using time stretch imaging, we demonstrated high-throughput image-based screening of budding yeast and rare breast cancer cells in blood with an unprecedented throughput of 100,000 particles/s and a record false positive rate of one in a million [30]. Our first rare cancer cell detection method was based on imaging metal beads conjugated to cells expressing specific surface antigens [30]. However, when downstream operations such as DNA sequencing and subpopulation regrowth are desired, the negative impacts of biomarkers on cellular behavior are often unacceptable.

2.3 Label-Free Phenotypic Screening

Phenotypic screening has been the basis for the discovery of new drugs and has also been widely used in biological research. High-throughput label-free cellular imaging leads to large scale and high dimensional phenotyping of cells in their natural conditions without biomarkers, enabling applications in circulating tumor cell detection when certain biomarkers are absent [31], as well as downstream analysis for studying the stochasticity in gene expression [32].

Imaging flow cytometry overcomes the throughput bottleneck in microscopic imaging as well as the low-content issue in conventional flow cytometry, making it a perfect candidate for label-free phenotypic screening. When combined with image analysis [33], it enables recognizing and quantifying multiple informative measures of cells, including morphology, protein localization, cell cycles, DNA content, etc.

2.4 Warped Time Stretch for Data Compression

Using warped group delay dispersion, it has been shown that one can reshape the spectro-temporal profile of optical signals such that signal intensity's time-bandwidth product is compressed [22–26]. The compression is achieved through time stretch dispersive Fourier transform in which the transformation is intentionally warped using an engineered group delay dispersion profile. This operation causes a frequency-dependent reshaping of the input waveform. Reconstruction (decoding) method depends on whether the information is in the spectral domain amplitude, or in the complex spectrum. In the time stretch camera, the image is encoded into the amplitude of the spectrum of a broadband optical pulse, and reconstruction consists of a simple nonuniform time-to-frequency mapping using the inverse of the warped group delay function.

To illustrate the concept in the context of time stretch imaging, we can consider a microscopic field of view consisting of a cell against a background such as a flow channel or a microscope slide (Fig. 2.2). In the time stretch imaging, the object is illuminated by an optical pulse that is diffracted into a 1-D rainbow. This maps the 1-D space into the optical spectrum. The spectrum is then linearly mapped into time using a dispersive optical fiber with a linear group delay. The mapping process from space to frequency to time is shown in Fig. 2.2a. The linearly stretched temporal waveform is then sampled by a digitizer resulting in uniform spatial sampling. This uniform sampling generates redundant data by oversampling the sparse peripheral sections of the field of view. Such a situation evokes comparison to the mammalian eye where central vision requires high resolution while coarse resolution can be tolerated in the peripheral vision. In the eye, this problem is solved through nonuniform photoreceptor density in the retina. The Fovea section of the retina has a much higher density of photoreceptors than the rest of the retina and is responsible for the high resolution of central vision.

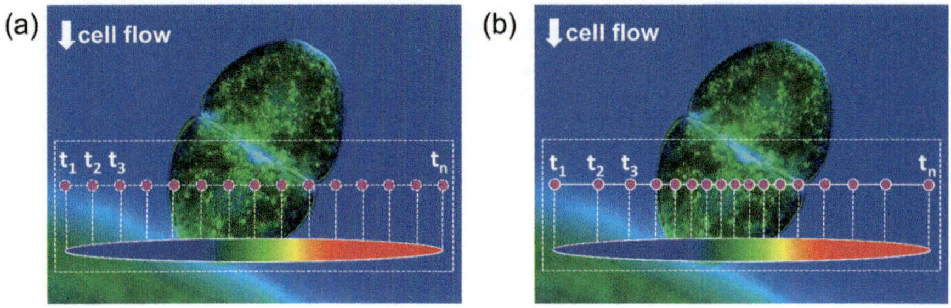

Fig. 2.2 Illustration of warped-stretch transform in imaging. (**a**) In conventional time stretch imaging (STEAM), the spectrum-to-time mapping is uniform, and the pixels are assigned equally distanced across the field of view. (**b**) Using a nonlinear group delay profile in the spectrum-to-time mapping process results in a nonuniform sampling of the line image, assigning more pixels to the information-rich central part of field-of-view and less to the low-entropy peripherals

We solve this problem by nonuniform mapping of spectrum into time via a warped group delay. The warped (anamorphic) space to frequency to time mapping is illustrated in the dotted box in Fig. 2.2b. After uniform sampling in time, this leads to higher sampling density in the central region of the field of view and lower density in the sparse peripheral regions. The reconstruction is a simple unwarping using the inverse of the group delay.

Part II
Inspection and Vision

Chapter 3
Nanometer-Resolved Imaging Vibrometer

Conventional laser vibrometers are incapable of performing multi-dimensional vibrometry at high speeds because they build on single-point measurements and rely on beam scanning, significantly limiting their utility and precision. Here we introduce a laser vibrometer that performs high-speed multi-dimensional imaging-based vibration and velocity measurements with nanometer-scale axial resolution without the need for beam scanning. As a proof-of-concept, we demonstrate real-time microscopic imaging of acoustic vibrations with 1 nm axial resolution, 1200 image pixels, and 30 ps dwell time at 36.7 MHz scan rate.

3.1 Introduction

Laser vibrometry is a powerful tool for measuring surface vibrations and displacements in a noncontact and noninvasive manner. It has been used in a diverse range of scientific [34, 35], industrial [34–37], and biomedical [34, 36, 38, 39] applications. Common industrial applications include nondestructive inspection and diagnosis of aircraft components, musical instruments, hard disk drives, microelectromechanical systems (MEMS), and automotive brakes [34–36]. Furthermore, laser vibrometers are widely employed in biological research and clinical environments for diagnosis of tympanic membranes [38, 39], observation of insect communication [34, 36], and evaluation of dental instruments [34, 36].

Unfortunately, conventional methods for laser vibrometry such as laser Doppler vibrometry [34–39] are unable to perform imaging based vibration measurements at high speeds. This is because their operation builds on single-point measurements and relies on beam scanning for multi-dimensional laser vibrometry. In other words, the scan rate of conventional multi-dimensional laser vibrometers (also called scanning vibrometers) is limited by that of laser scanners although the single-point measurement itself is fast (on the order of $\sim$10 MHz or higher). Currently,

A. Mahjoubfar et al., *Artificial Intelligence in Label-free Microscopy*,
DOI 10.1007/978-3-319-51448-2_3

the maximum scan rates provided by commercially available laser scanners (e.g., galvanometric mirrors [40] and acousto-optic deflectors [41]) are ~100 kHz in 1D line scans and ~1 kHz in two-dimensional (2D) raster or spiral scans. This speed limitation significantly restricts the utility and precision of laser vibrometers, especially in high-speed vibrometry applications including MEMS devices and impact analysis [34–36].

Efforts have been made to mitigate the speed limitation in multi-dimensional laser vibrometers. One of the popular methods is the illumination of the target with multiple laser beams [42, 43], but the number of image pixels is significantly limited (typically up to ~10) [42, 43] by the complexity and cost of the required optical components (e.g., multiple lasers, interferometers, and photodetectors). Another type of vibrometer that does not require beam scanning relies on the use of an array detector, i.e., the complementary metal–oxide–semiconductor (CMOS) camera [44], and hence its scan rate is limited by the frame rate of the camera (up to ~10 kHz) [44] and also the trade-off between the number of pixels and frame rate.

In this chapter, we propose and demonstrate a laser vibrometer that overcomes the limitations in the conventional multi-dimensional laser vibrometers and achieves high-speed imaging-based surface vibration measurements with nanometer-scale axial resolution at ~100 times higher scan rates than the conventional methods. This method is an extension of the recently developed ultrafast imaging technology known as serial time-encoded amplified imaging/microscopy (STEAM) [10, 45, 46] to depth-resolved multi-dimensional imaging. By stretching in time a spectrally coded image, this method does not require beam scanning for multi-dimensional vibrometry. Furthermore, the superior temporal resolution of this method also enables multi-dimensional velocimetry as the velocity of the surface can be obtained from the axial position of the surface. The method's fast shutter speed (dwell time) ensures nearly instantaneous frame acquisition and eliminates image blurring. As a proof-of-concept, we demonstrate real-time depth-resolved imaging of acoustic vibrations up to 30 kHz with 1 nm axial resolution, 1200 image pixels, and 30 ps dwell time at 36.7 MHz scan rate.

3.2 Experimental Demonstration

An experimental apparatus of the proposed method, which we refer to as STEAM vibrometry, is shown in Fig. 3.1. The optical source is a mode-locked femtosecond pulse fiber laser with a pulse repetition rate of 36.7 MHz. After supercontinuum generation in a highly nonlinear fiber and band-pass filtering, a nearly flat spectral shape with ~20 nm bandwidth centered at 1590 nm is produced for target illumination. A pair of diffraction gratings with 1100 lines/mm spatially disperses the pulses along a 1D line, which are directed toward the vibrating target. The reflected pulses are interfered with the reference pulses in a Michelson interferometer, resulting in the spectral interference between the test and reference pulses. Here the lateral

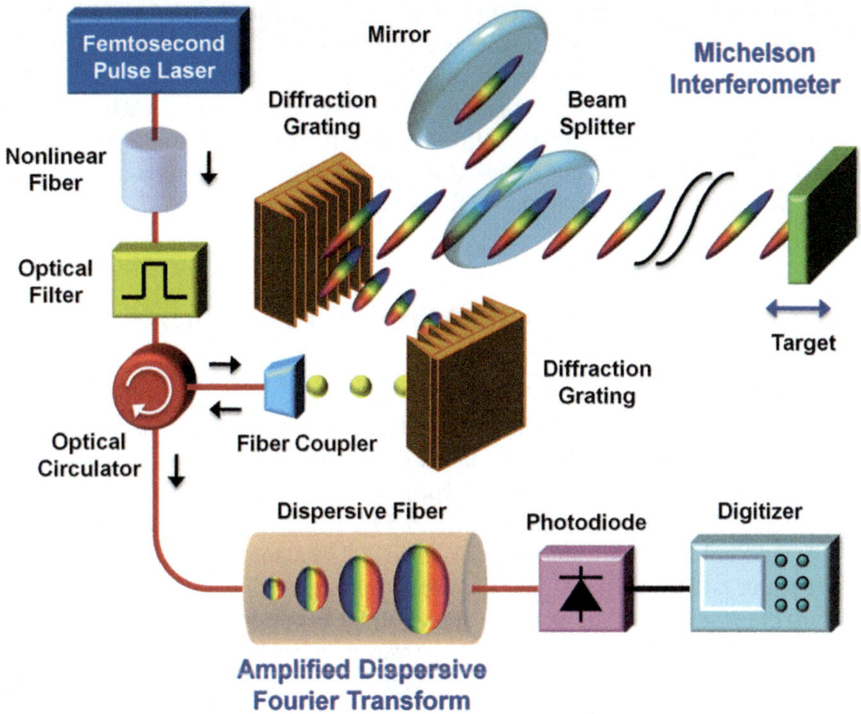

Fig. 3.1 Schematic of the STEAM vibrometer. The principle of the method is three-fold: (1) encoding of the lateral and axial coordinates of the target into the different frequencies and corresponding amplitudes of a spatially dispersed broadband pulse which spectrally interferes with a reference pulse, (2) amplified dispersive Fourier transformation in which the spectrum is mapped into a temporal waveform, time stretched so that it can be digitized in real time, and simultaneously amplified in the optical domain, and (3) Hilbert transformation on the detected pulse in the digital domain to extract the axial information of the target

and axial coordinates of the target are encoded into the different frequencies and corresponding amplitudes of each back-reflected spatially dispersed pulse, respectively. This situation may be better understood by interpreting the optical configuration in such a way that multiple continuous-wave lasers are incident onto different spatial coordinates of the target in a shared Michelson interferometer with their longitudinal modes locked.

The interferometrically combined pulses return to the same optics, but are directed via an optical circulator toward the amplified dispersive Fourier transformer (ADFT) [10, 45, 47] in which a dispersive fiber with -1200 ps/nm dispersion is optically pumped by four continuous-wave lasers with $\sim$100 mW of optical power at 1470, 1480, 1480, and 1490 nm for distributed Raman amplification. In the dispersive medium, the spectrum of each interfered pulse is stretched and converted into an amplified temporal waveform. This ADFT process is critical for high-speed laser vibrometry because the optical amplification before photon-to-electron conversion overcomes the fundamental trade-off between sensitivity and speed [10, 47]. The pulses are captured by a high-speed photodiode with 15 GHz bandwidth and digitized by a real-time oscilloscope with 16 GHz bandwidth and 50 GS/s sampling rate. Hilbert transformation is applied in the digital domain to each spectrally interfered pulse to obtain the axial information of the target at multiple points along the 1D line. Each pulse acquires one scan and the pulse repetition rate corresponds to the scan rate (frame rate) of the STEAM vibrometer.

3.3 Theoretical Study of the Vibrometer Performance

The basic capabilities of the STEAM vibrometer (i.e., image pixel number, axial resolution, and dwell time) can be estimated from the parameters of its components. First, the number of image pixels on the target (N) is found from the total dispersion in the dispersive fiber ($D = -1200$ ps/nm), the optical bandwidth ($\Delta\lambda = 20$ nm), and the sampling rate of the digitizer ($f_{\text{dig}} = 50$ GS/s) to be $N = |D| \cdot \Delta\lambda \cdot f_{\text{dig}} = 1200$ while the number of resolvable points is about 200 from the spectral resolution of the ADFT process [18]. Second, the axial resolution is given by the dynamic range (bit depth) of the digitizer. The axial resolution (Δz) can be found from the expression, $0.5 \sin(2 \cdot k \cdot \Delta z) = 2^{-n}$, where k is the wavenumber [$k = 2\pi/(1590\,\text{nm})$] and n is the bit depth of the digitizer ($n = 8$ bits), to be $\Delta z = 0.99$ nm. Finally, the dwell time is estimated from the bandwidth of each subpulse (20 nm/$\sim$200) and the time-bandwidth product to be $\sim$30 ps (assuming that the subpulses are transform limited).

3.4 Experimental Results

We evaluated the basic performance of the STEAM vibrometer. In Fig. 3.2a, the temporal waveform of a single interfered pulse captured by the photodiode is compared with the optical spectrum measured by a conventional optical spectrum analyzer. This verifies the equivalence of the two waveforms and hence validates the STEAM vibrometer. As shown in Fig. 3.2b, repetitive pulses (scans) detected by the photodiode indicate that the STEAM vibrometer operates at 36.7 MHz scan rate.

To show the utility of the STEAM vibrometer, we monitored the performance of an acoustic speaker. For better sensitivity, a thin reflective plate was attached to the diaphragm of the acoustic speaker. The speaker was driven up to 30 kHz (nearly its upper frequency limit). Figure 3.3 shows the 30 kHz surface vibration of the diaphragm captured by the STEAM vibrometer with $\sim$1 nm axial resolution (which agrees with our estimated axial resolution of 0.99 nm).

In addition to the amplitude of the surface vibration, we also obtained the velocity of the diaphragm from the axial coordinates of the surface as shown in Fig. 3.4. The Doppler frequency shift in the frequency comb lines caused by the acoustic vibration ($\sim$830 Hz frequency shift) is negligible.

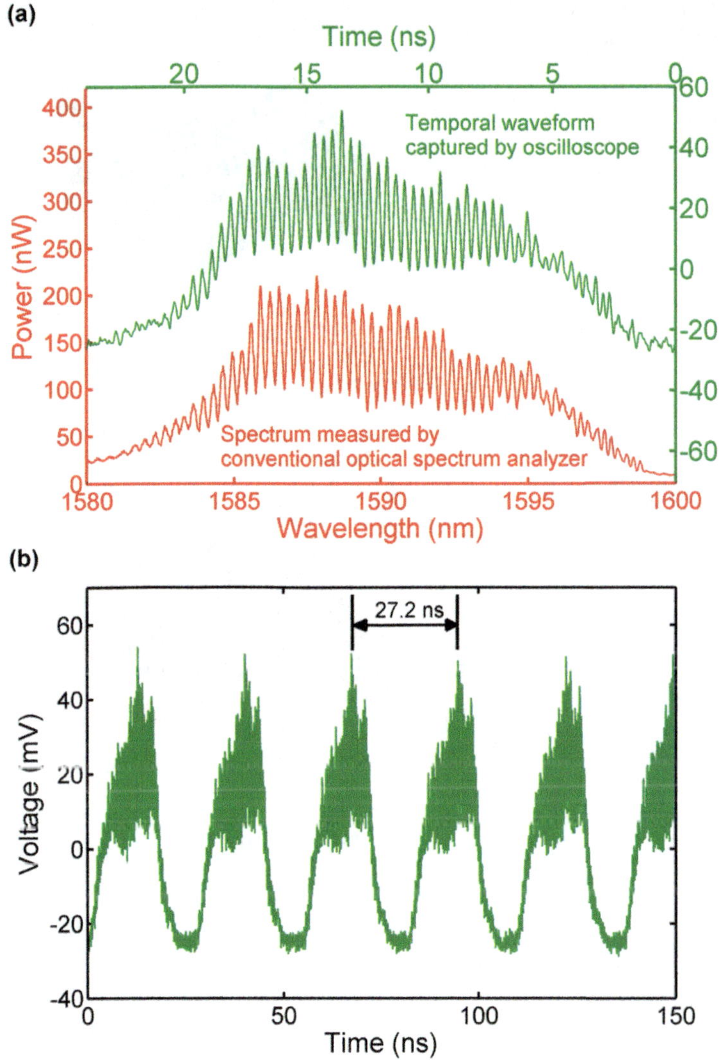

Fig. 3.2 Basic performance of the STEAM vibrometer. (**a**) Temporal waveform of a single interfered pulse captured by the photodiode in comparison with the optical spectrum measured by a conventional optical spectrum analyzer. (**b**) Repetitive pulses (scans) with a time interval of 27.2 ns detected by the photodiode indicating that the STEAM vibrometer operates at 36.7 MHz scan rate

3.5 Conclusion

In summary, we proposed and demonstrated an optical system that performs high-speed multi-dimensional imaging-based vibrometry and velocimetry with nanometer-scale axial resolution without the need for beam scanning. As a

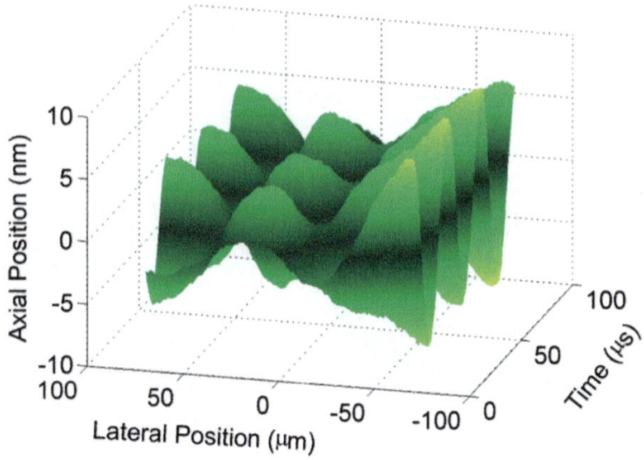

Fig. 3.3 Surface vibration of the acoustic diaphragm captured by the STEAM vibrometer with ∼1 nm axial resolution and ∼30 ps dwell time. The diaphragm was driven to vibrate at 30 kHz

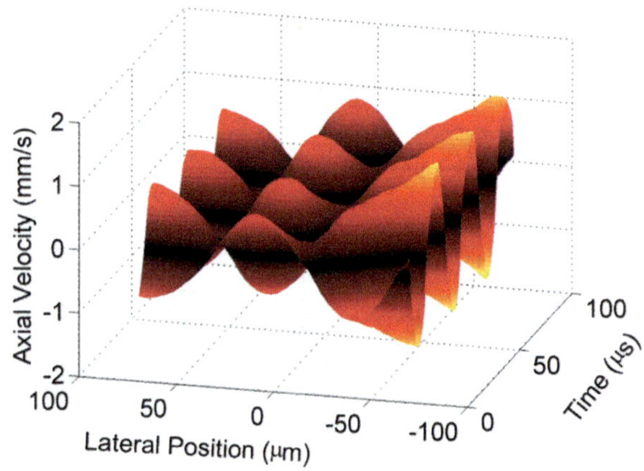

Fig. 3.4 Axial velocity of the acoustic diaphragm obtained by the STEAM vibrometer. The diaphragm was driven to vibrate at 30 kHz (the same as in Fig. 3.3)

proof-of-concept, we showed real-time 1D imaging of fast acoustic vibrations with 1 nm axial resolution, 1200 image pixels, and 30 ps dwell time at 36.7 MHz scan rate. While we performed 1D cross-sectional imaging in this proof-of-principle demonstration, the technique can naturally be extended to 2D by using a 2D spatial disperser [10, 48].

Chapter 4
Three-Dimensional Ultrafast Laser Scanner

Laser scanners are essential for scientific research, manufacturing, defense, and medical practice. Unfortunately, often times the speed of conventional laser scanners (e.g., galvanometric mirrors and acousto-optic deflectors) falls short for many applications, resulting in motion blur and failure to capture fast transient information. Here, we present a novel type of laser scanner that offers roughly three orders of magnitude higher scan rates than conventional methods. Our laser scanner, which we refer to as the hybrid dispersion laser scanner, performs inertia-free laser scanning by dispersing a train of broadband pulses both temporally and spatially. More specifically, each broadband pulse is temporally processed by time stretch dispersive Fourier transform and further dispersed into space by one or more diffractive elements such as prisms and gratings. As a proof-of-principle demonstration, we perform 1D line scans at a record high scan rate of 91 MHz and 2D raster scans and 3D volumetric scans at an unprecedented scan rate of 105 kHz. The method holds promise for a broad range of scientific, industrial, and biomedical applications. To show the utility of our method, we demonstrate imaging, nanometer-resolved surface vibrometry, and high-precision flow cytometry with real-time throughput that conventional laser scanners cannot offer due to their low scan rates.

4.1 Introduction

High-speed multidimensional laser scanning technology has numerous applications in research [44, 49–55], manufacturing [49–51, 56–60], defense [49, 50, 56, 57, 60], and biomedicine [44, 49, 52–54, 61–63] for sensing and imaging of moving objects and dynamic processes. Low scan rates cause motion blur in images or missing fast

A. Mahjoubfar et al., *Artificial Intelligence in Label-free Microscopy*,
DOI 10.1007/978-3-319-51448-2_4

transient phenomena in sensing. Also, high-speed scanning capability is needed in high-throughput analysis of a large number of objects or a wide field of view in a reasonable duration of time [12, 49–51, 55, 60–63].

Various types of laser scanners have been developed over the past few decades. The most commonly used type of laser scanners including MEMS scanners [40] is based on beam steering by galvanometric mirrors. However, their linear scan rates because of inertia are limited to about 10 kHz. If two of these scanners are aggregated to perform 2-dimensional (2D) raster scans, the overall raster scan rate is limited to about 100 Hz. Another type of laser scanner is based on diverting laser beams by acousto-optic deflectors (AODs). They are about one order of magnitude faster than galvanometric mirror scanners in both linear and 2D raster scans [41, 49]. Finally, a combination of a frequency-tunable laser and diffractive optics can be used to form a laser scanner at scan rates comparable to AODs [64, 65].

Recently, we have demonstrated a new type of inertia-free ultra-fast laser scanner that can achieve about three orders of magnitude faster scan rates than the conventional methods [13]. The operation principle of this method, namely the hybrid dispersion laser scanner (HDLS), is based on probing different points of a target with frequency components of a linearly chirped broadband optical pulse at different times. In this chapter, we present results from our demonstration of linear scans at 90.8 MHz, 2D raster scans at 105.4 kHz, and 3D scanning surface vibrometry with nanometer axial resolution.

4.2 Principle of Hybrid Dispersion Laser Scanner

The concept of HDLS relies on the transformation from spectral to temporal and spatial domains, respectively (Fig. 4.1). First, by a process called dispersive Fourier transformation [10, 45, 47, 66, 67] based on group-velocity dispersion, the spectra of broadband optical pulses of a mode-locked laser are mapped into temporal waveforms. Then, a spatial dispersive element such as a diffraction grating or a virtually imaged phased array (VIPA) maps the spectrum of chirped pulses onto a line over the object such that different wavelength components hit the target at different positions and times. The reflected or scattered light from the target is then detected by a single-pixel photodetector. Wavelength components of each laser pulse perform one linear scan, and therefore, the scan rate is same as the repetition rate of the mode-locked laser. A complementary scanner for other axis can be added to achieve 2D raster scans with HDLS.

Based on the HDLS concept described, we designed and implemented a multi-dimensional laser scanner in the industrially and biomedically important spectral range of 800 nm (Fig. 4.2). A Ti:Sapphire femtosecond mode-locked laser with a repetition rate of 90.8 MHz generates a train of broadband optical pulses centered at 814 nm. The process of wavelength-to-time mapping is performed with two pairs of prisms followed by a dispersive fiber. Pulses are then collimated into free space and

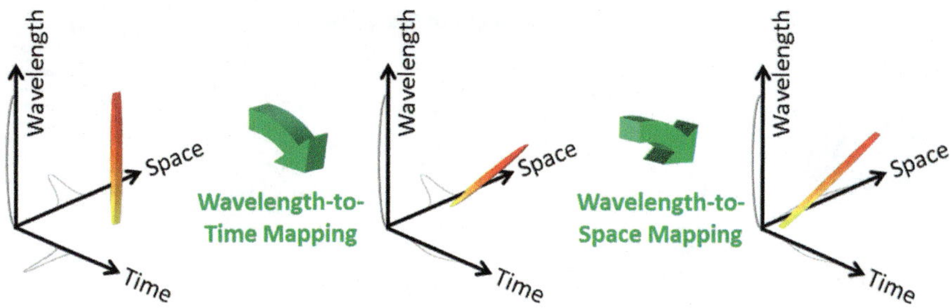

Fig. 4.1 Concept of HDLS. HDLS operation relies on two high-speed mapping processes. First, the spectrum of each broadband optical pulse generated by a mode-locked laser is mapped to time. This wavelength-to-time mapping is performed by a temporal dispersive element such as a dispersive optical fiber or prism pair. Next, wavelength-to-space transformation is used to direct each wavelength component of the optical pulse to a unique point on the target. Overall, a one-to-one mapping between time and space is formed. Therefore, each point in HDLS' field of view is sampled with an individual wavelength component of the optical pulse at a specific time. The repetition rate of the mode-locked laser determines the sampling rate of the HDLS

scanned in the vertical direction by an acousto-optic deflector at 105.4 kHz. A pair of diffraction gratings performs the wavelength-to-space mapping, which is the key to fast scanning capability of HDLS at 90.8 MHz in the horizontal direction.

Different wavelength components of each laser pulse hit the target at different times, such that a single-pixel photodetector can be used to measure their reflections. The electrical signal of the photodetector corresponding to the waveform of the reflected optical pulses is captured by a high-speed digitizer (50 GS/s, 20 GHz bandwidth oscilloscope) (Fig. 4.3a). Digital waveforms are processed and combined in Matlab to generate multi-dimensional scan profiles. To validate the wavelength-to-time mapping implemented by the prism pairs and dispersive fiber, the spectrum of the reflected pulses from a fixed target is measured with a conventional spectrum analyzer and compared to the waveforms captured by the oscilloscope (Fig. 4.3b). Good agreement between them confirms that we can measure the spectral information of laser pulses at the pulse repetition rate of the mode-locked laser that is well beyond the scan rate of conventional spectrum analyzers.

4.3 Applications of Hybrid Dispersion Laser Scanner

In order to visualize the operation of the HDLS, we scanned a high-reflective substrate with letters "UCLA" engraved on it, and compared the results side-by-side with an image taken by a regular CCD camera (Fig. 4.4). For this experiment, the target was 90° rotated around the illumination axis with respect to the images shown, so that the vertical scans in the image are performed by the HDLS while horizontal scans in the image are implemented by the AOD.

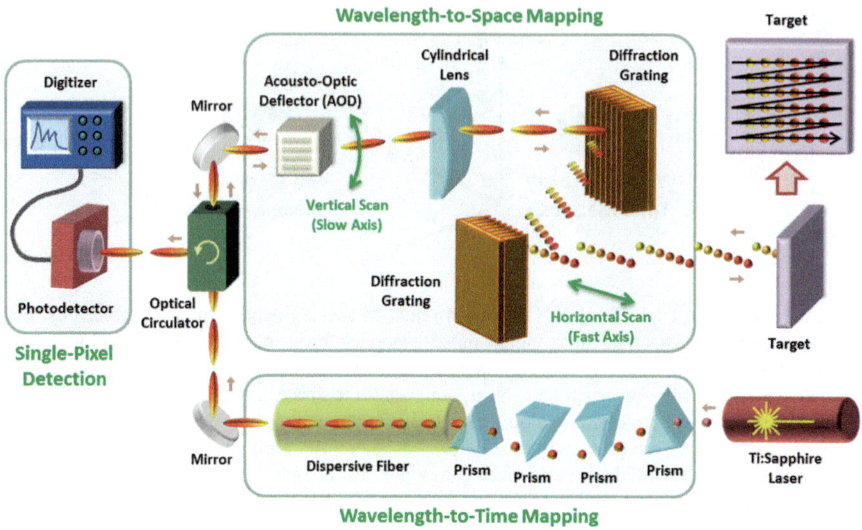

Fig. 4.2 HDLS experimental setup. In a 2D demonstration of laser scanning with the HDLS, optical pulses generated by a mode-locked Ti:Sapphire laser with a center wavelength of 814 nm and a repetition rate of 90.8 MHz are dispersed in time using two pairs of prisms and a dispersive fiber. Pulses are deflected in the vertical direction using an acousto-optic deflector (AOD) at 105.4 kHz. Subsequently, the spectrum of each pulse is mapped onto a horizontal line using a pair of diffraction gratings. A combination of the vertical deflection and horizontal mapping leads to a 2D raster scan on the target. The pulse reflection off the surface of the target is converted via an optical circulator to an electrical signal using a single-pixel high-speed photodetector. This is possible due to the prior wavelength-to-time mapping, so that each wavelength component reaches the photodetector at a unique time, and the information of different points of the target are not overlapped. A 50 GS/s digitizer acquires the electrical signal from the photodetector, which corresponds to the spectrum of the optical pulses. After correction for the background envelope, the spectrum of each pulse reveals one horizontal line scan image of the target. Stacking up many of these line scans in accordance with the AOD scan frequency leads to a 2D raster scan of the target

Combining the HDLS with an interferometer, we performed 3D surface profilometry or 2D surface vibrometry (Fig. 4.5). Here we used a Michelson interferometer to encode phase delays of different points on the target into wavelength components of the illumination pulses. The interferograms are then captured in time and analyzed offline by Hilbert transformation to extract the phase variations, which correspond to the axial positions. Our experimental setup enables an axial resolution of 0.4 nm at a scan rate of 105.4 kHz. As an illustrative demonstration, we captured vibrations of a reflective diaphragm oscillating at 1 kHz (Fig. 4.6).

Finally, as an example of the HDLS' biomedical utility, we demonstrated high-precision high-throughput flow cytometry using the HDLS. Low spatial resolution of conventional flow cytometers causes a considerable number of false positive events that result in statistical error in subpopulation analysis. For instance, they are not effective for detection of multiple cells (i.e., doublets, triplets, etc.) (Fig. 4.7a).

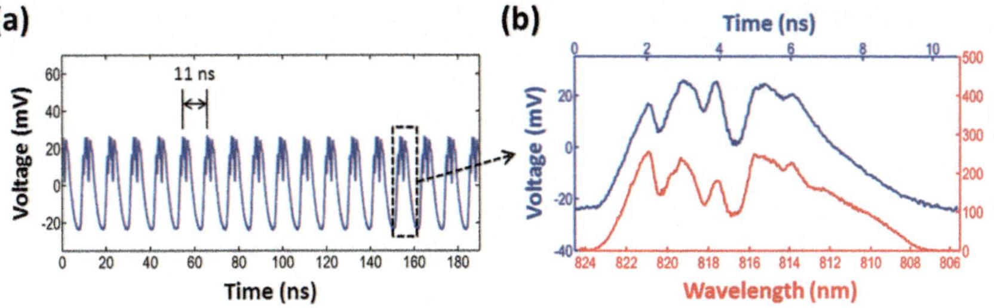

Fig. 4.3 Wavelength-to-time mapping with dispersive Fourier transformation. (**a**) Optical pulses reflected off the target corresponding to horizontal line scans at different deflection angles of the AOD are measured by a high-speed photodetector. The period of the horizontal scans is about 11 ns, which corresponds to the mode-locked laser's pulse repetition rate (90.8 MHz). (**b**) Good agreement between the amplitude of the photodetector signal measured with the digitizer (shown in *blue*) and the power spectrum measured with a conventional optical spectrum analyzer indicates the demonstration of the wavelength-to-time mapping using the prism pairs and dispersive fiber in the 800 nm band

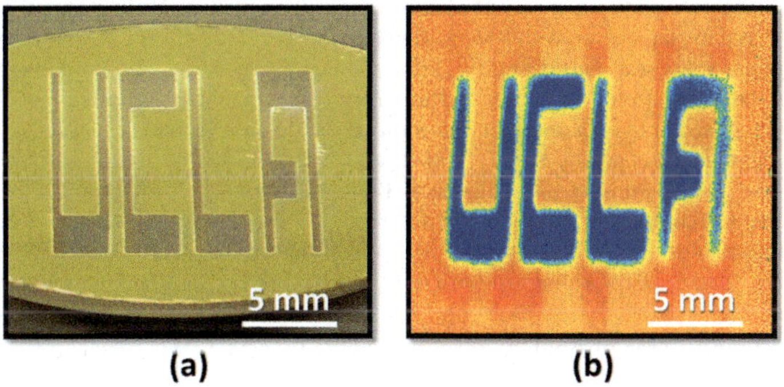

Fig. 4.4 Imaging with the HDLS. (**a**) Image of the word "UCLA" engraved on the surface of a reflective substrate captured by a CCD camera. (**b**) Image of the same sample captured by the HDLS. The word "UCLA" is clearly shown

We used inertial focusing microfluidic technology [68] to precisely align otherwise randomly positioned cells in a single stream with no need for sheath flow (Fig. 4.7b). The microfluidic channel is custom-made on a substrate dielectric mirror from thermoset polyester (TPE) for stability, robustness, and increased precision of cell focusing. HDLS pulses scan the stream of cells, and the forward scattering is reflected back by the substrate mirror for measurement.

We tested the performance of the HDLS and conventional flow cytometer for size-based identification of white blood cells and MCF7 breast cancer cells (Fig. 4.8a). Since the HDLS flow cytometer is more precise in distinguishing multiple cells, e.g., doublets, the count of white blood cells that have unusually larger size is reduced, and therefore, the false positive rate decreases. This improvement in size-based classification of MCF7 breast cancer cells from white blood cells is more evident in comparison of receiver operating characteristic (ROC) curves of

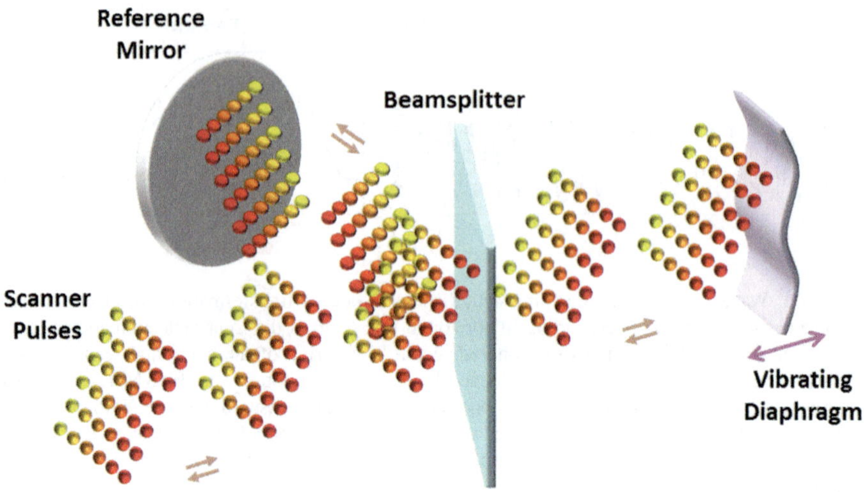

Fig. 4.5 3D surface profilometry or 2D surface vibrometry with the HDLS. 2D raster scans by the HDLS are used in conjunction with a Michelson interferometer to perform 2D surface vibrometry. A beamsplitter splits scan pulses into two arms. Optical pulses in one arm hit the target, and the light in the other arm (reference arm) is reflected intactly by a mirror. Reflected pulses from both arms are combined at the beamsplitter and form an interference pattern. If the reflectivity of the vibrating target is not changing rapidly, Hilbert transformation can be used to extract the relative optical phase of each wavelength component. Therefore, variations of the optical path length at each wavelength component are measured and used to form 3D surface profiles of the vibrating sample at a scan rate of 105.4 kHz with 0.4 nm axial resolution

the HDLS and conventional flow cytometer (Fig. 4.8b). We observed that for the same specificity, the sensitivity of our method is significantly higher than that of the conventional flow cytometer.

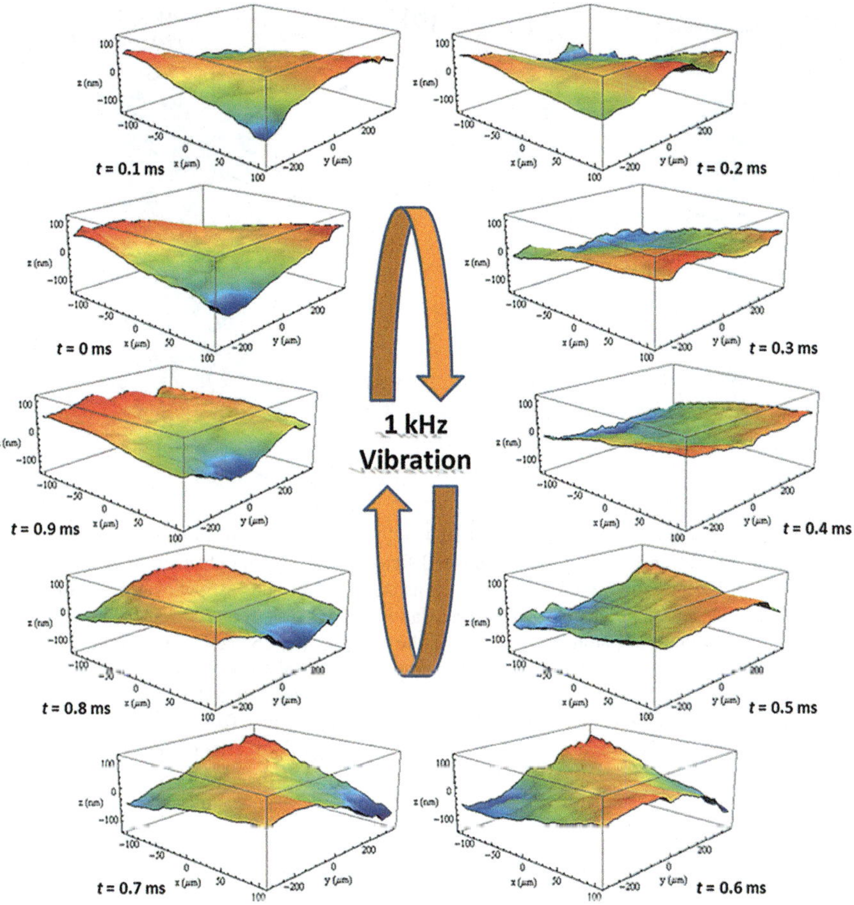

Fig. 4.6 Surface vibration captured by the HDLS. Frames from 3D scans of a vibrating diaphragm by the HDLS show a period of nanomechanical vibrations at 1 kHz. Only one of every ten scans is shown here

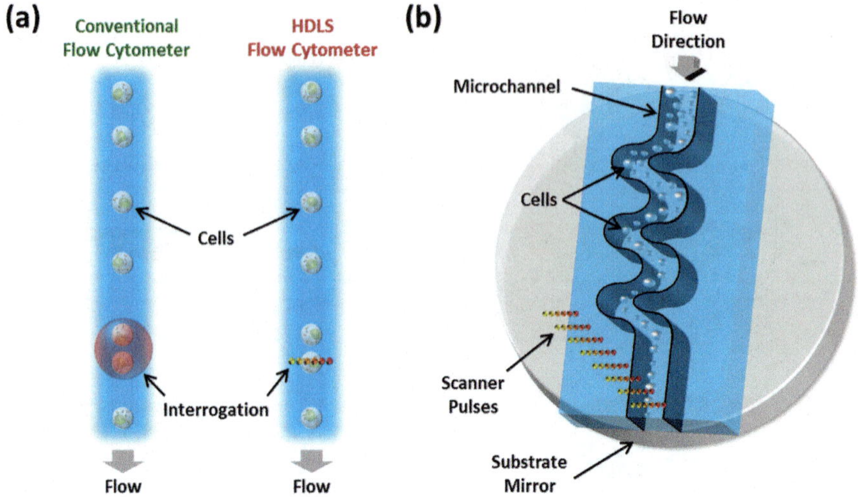

Fig. 4.7 Comparison of conventional and HDLS flow cytometers. (**a**) In regular flow cytometry, a single interrogation beam covers the desired field of view in the channel. Therefore, it does not efficiently differentiate multiple cells such as doublets. In HDLS flow-cytometer, diffraction limited wavelength components of the interrogation beam cover the required field of view, and extract high-resolution spatial information of the sample. HDLS data can be used to identify abnormalities, e.g., multiple cells and result in a lower statistical error. (**b**) In our demonstration of HDLS flow cytometer, an inertial focusing microfluidic channel with a dielectric mirror substrate is used to order randomly distributed cells into a single stream. The microfluidic device was fabricated using standard replica molding methods in thermoset polyester (TPE) to ensure stability. HDLS scan pulses are focused on the stream. Forward-scattered light from the cells is reflected by substrate mirror and collected by an objective lens

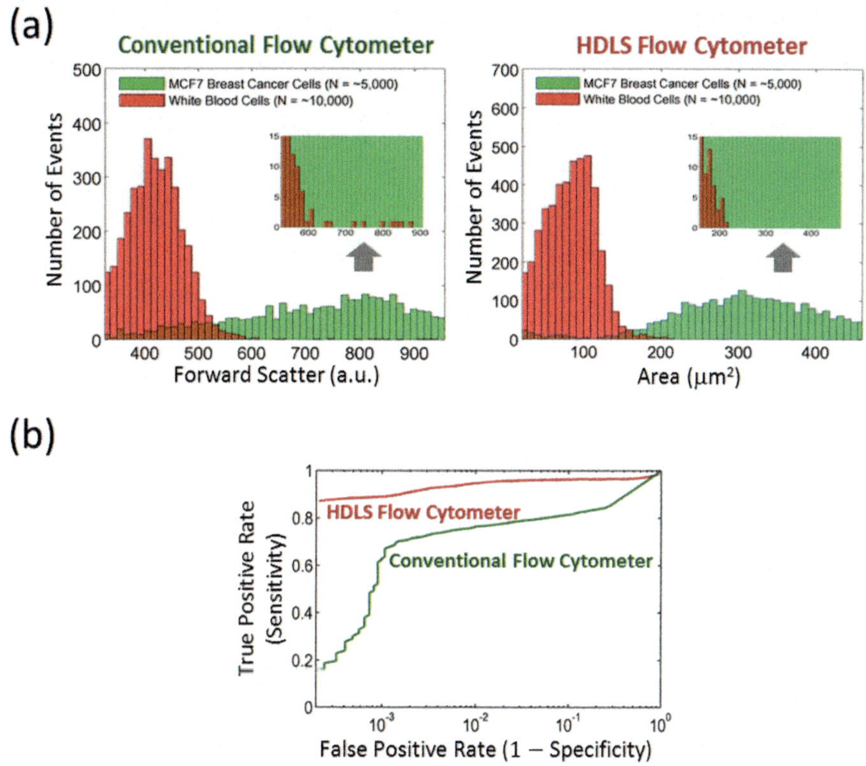

Fig. 4.8 Experimental results of HDLS flow cytometer. (**a**) Identical samples of white blood cells and MCF7 breast cancer cells are measured separately with conventional and HDLS flow cytometers. There is a considerable overlap in forward scattering range of these cell types for a conventional flow cytometer. However, this overlap decreases significantly for HDLS flow cytometer measurements because white blood cell multiples are not identified as MCF7 cancer cells. (**b**) Receiver operating characteristic (ROC) curves based on identification of white blood cells and MCF7 breast cancer cells show that without sacrificing throughput, HDLS flow cytometer achieves higher specificity and sensitivity than a conventional flow cytometer

Part III
Biomedical Applications

Chapter 5
Label-Free High-Throughput Phenotypic Screening

Flow cytometry is a powerful tool for cell counting and biomarker detection in biotechnology and medicine especially with regards to blood analysis. Standard flow cytometers perform cell type classification both by estimating size and granularity of cells using forward- and side-scattered light signals and through the collection of emission spectra of fluorescently labeled cells. However, cell surface labeling as a means of marking cells is often undesirable as many reagents negatively impact cellular viability or provide activating/inhibitory signals, which can alter the behavior of the desired cellular subtypes for downstream applications or analysis. To eliminate the need for labeling, we introduce a label-free imaging-based flow cytometer that measures size and cell protein concentration simultaneously either as a stand-alone instrument or as an add-on to conventional flow cytometers. Cell protein concentration adds a parameter to cell classification, which improves the specificity and sensitivity of flow cytometers without the requirement of cell labeling. This system uses coherent dispersive Fourier transform to perform phase imaging at flow speeds as high as a few meters per second.

5.1 Introduction

Cell protein content measurement can be used in many biomedical applications such as blood doping detection [69], infection monitoring [70], drug development and screening [71], studies of necrosis and apoptosis [72, 73], cell cycle progression and differentiation [74–76], and in cancer diagnostics [77–79]. Current methods for cell protein concentration measurement include electrical methods based on dielectrophoresis [80], mechanical methods based on microchannel cantilevers [69], and optical methods based on scattering patterns [81], emission spectra of external cavity lasers [82], and holographic and phase microscopy [83–86]. These methods are either inherently too slow for high-speed flow cytometry applications or require

A. Mahjoubfar et al., *Artificial Intelligence in Label-free Microscopy*,
DOI 10.1007/978-3-319-51448-2_5

feedback mechanisms [44] to provide necessary precision. Furthermore, size-based classification can also be used for label-free identification of cells of interest in a suspension stream [87]. However, due to significant overlap of size ranges between most mammalian cells, size-based technologies require additional layers of parametric gating to be useful as a diagnostic tool [88]. It is known that the refractive index of a cell is proportional to its protein content [89]. As such, the simultaneous measurement of refractive index and size of cells would be predicted to provide two independent parameters for cell classification.

In this chapter, we propose a fast and high-precision optical cell density and size measurement method based on serial time-encoded amplified microscopy (STEAM) [10]. STEAM is a continuous imaging technique that captures tens of million frames-per-second with sub-nanosecond shutter speed. However, earlier versions of STEAM were dependent on cell labeling due to low intensity contrast of individual cells [30]. Here, we introduce a new configuration of STEAM capable of high-speed phase microscopy and demonstrating label-free single-cell classification and diagnostics. In addition, we demonstrate a new design of STEAM that minimizes loss and chromatic aberration, decreases polarization sensitivity, and results in a smaller footprint [90]. In contrast to previous implementations of STEAM, which were based on refractive optics, the new design employs reflective optics.

The basic principle of STEAM involves two steps both performed optically. In the first step, the spectrum of a broadband optical pulse is converted by a spatial disperser into a rainbow that illuminates the target. Therefore, the spatial information (image) of the object is encoded into the spectrum of the resultant reflected or transmitted rainbow pulse. A 1D rainbow is used in flow imaging as the flow causes the cell to be scanned in the second dimension. In the second step, the spectrum of the image-encoded pulse is mapped into a serial temporal signal that is stretched in time to slow it down such that it can be digitized in real-time [9]. This optically amplified time stretched serial stream is detected by a single-pixel photodetector and the image is reconstructed in the digital domain. Subsequent pulses capture repetitive frames, hence the laser pulse repetition rate corresponds to the frame rate of STEAM and the shutter speed (exposure time) corresponds to the temporal width of the pulse. The key innovations in STEAM that enable high-speed real-time imaging are photonic time stretch for digitizing fast images in real-time and the optical image amplification for compensating the low number of photons collected during the ultra-short shutter time.

The contrast limitations of label-free single-cell imaging of STEAM led us to develop a derivative of STEAM, referred to as Coherent-STEAM, to capture phase images of cells in flow. We use a Michelson interferometer to map the phase image of cells into the spectrum of broadband optical pulses. This phase imaging technique exploits the fast shutter speed of STEAM to freeze path length fluctuations of interferometer arms and attains nanometer phase resolution with no need for feedback stabilization of the interferometer [12, 13, 16]. We use Coherent-STEAM to measure the refractive index of individual cells in an imaging flow cytometer by simultaneous measurement of size and total optical phase-shift induced by the cells. As an example, we use our label-free STEAM-based cell classifier to distinguish

OT-II T cell hybridoma from SW480 epithelium cancer cells. We show that adding protein concentration to size as an additional classification parameter increases accuracy and specificity in flow cytometry.

5.2 Experimental Setup

A mode-locked fiber laser generates pulses at 1565 nm with a repetition rate of 36.128 MHz and a pulse width slightly less than 100 fs (Fig. 5.1). Pulses are spectrally broadened with a highly nonlinear fiber to approximately 100 nm bandwidth [91]. A short dispersion compensating fiber with an overall dispersion of 60 ps/nm is used to temporally broaden pulses to 1.2 ns, so an erbium doped fiber amplifier (EDFA) can amplify them without any distortion. Amplified pulses enter a coarse wavelength division multiplexing (WDM) filter, and the output of 1591 nm channel is used to shape laser pulses with a considerably flat spectrum over 1581–1601 nm bandwidth. These pulses pass through an optical circulator and are coupled to free-space with a fiber collimator.

Free-space laser pulses are linearly polarized with quarter- and half-wave plates, and then they are spatially dispersed with a pair of reflection diffraction gratings, so that each wavelength component of the collimated beam is positioned at a different lateral point similar to a rainbow. A pair of 90° off-axis parabolic gold-coated mirrors with 152.4 and 25.4 mm reflected focal lengths are used to form a beam reducer that shrinks the rainbow beam six times. Parabolic gold-coated mirrors are used to minimize loss, aberration, and polarization sensitivity. In addition, a 15° off-axis parabolic gold-coated mirror with 635 mm reflected focal length and a 0.4 numerical aperture long working-distance objective lens further shrink the rainbow to about 130 μm field of view. Using reflective optics, we managed to improve the signal-to-noise ratio by about 9 dB. A beam splitter is used to form two arms of a Michelson interferometer. Different wavelength components of the rainbow are focused on a mirror in the reference arm and on the reflective substrate of a microfluidic device in the sample arm. Cells hydrodynamically focused at the center of the channel flow at a velocity of 1.3 m/s. The rainbow pulses pass through the cells and are reflected back by the mirror substrate of the microfluidic device. The total bandwidth of the pulses interrogating the cells in our Coherent STEAM is less than 20 nm centered at 1590 nm, giving a negligible fractional bandwidth of 1.3%. Therefore, the color-dependency of absorption is very small and can be easily neglected. The reflected pulses from the microfluidic device and reference mirror interfere at the beam splitter and return to the fiber, where they are directed with the optical circulator to an amplified time stretch system.

The amplified time stretch system is a combination of a Raman amplifier and a dispersive fiber to perform dispersive Fourier transform [9]. Four Raman pump lasers at 1450, 1470, 1490, and 1505 nm are used to amplify the signal for about 15 dB over the whole optical bandwidth uniformly. The dispersive fiber chirps and stretches each pulse in time to about 27 ns. So, different wavelength components

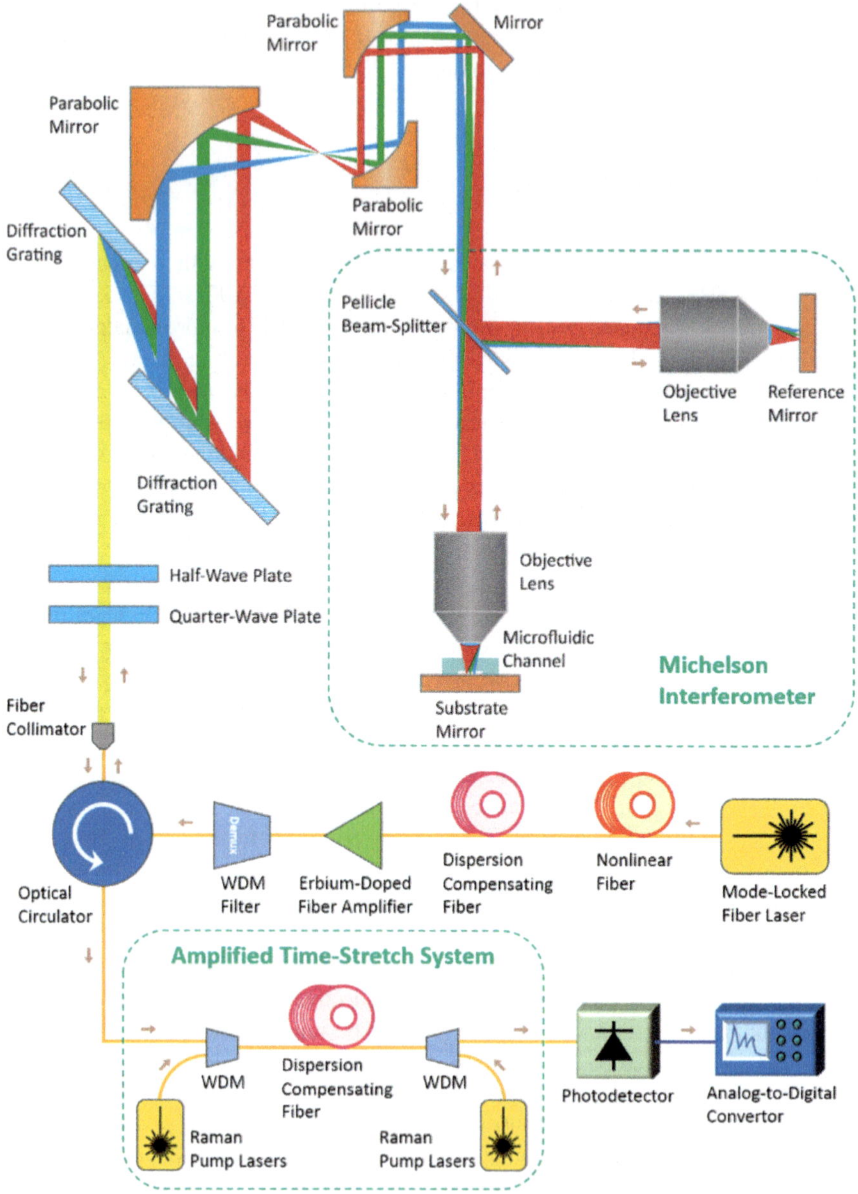

Fig. 5.1 Optical setup of Coherent-STEAM. A Coherent-STEAM setup is formed by combination of STEAM and a Michelson interferometer. A pair of diffraction gratings generates a 1D rainbow with different wavelength components imaging different points on the cells flowing in a microfluidic channel. A pellicle beam-splitter and two identical long working-distance objective lenses are used to form the interferometer for phase measurement. Back apertures of objective lenses are fully illuminated with each wavelength component of the broadband mode-locked laser pulses to ensure diffraction-limited resolution. An amplified time stretch system chirps, stretches, and amplifies each pulse, so that different wavelength components reach the photodetector serially. A very shallow microfluidic channel with hydrodynamic focusing is designed and fabricated to align cells within the focal depth of the system

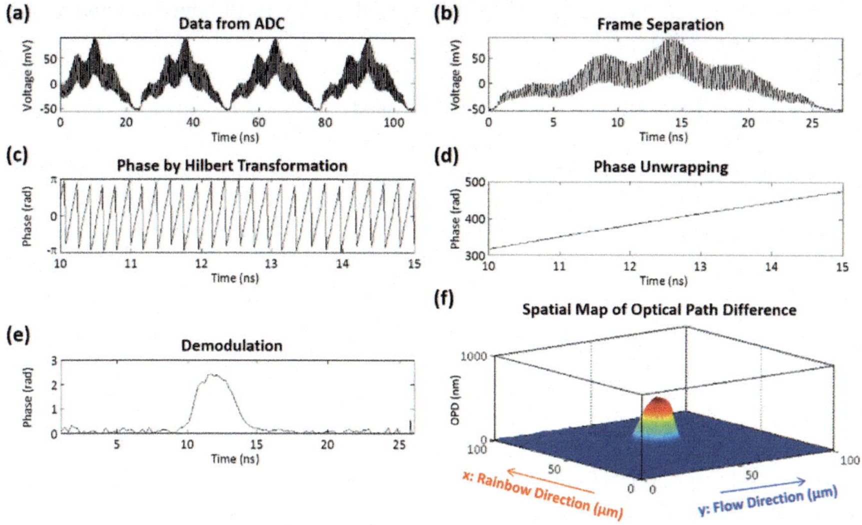

Fig. 5.2 Digital signal processing of Coherent-STEAM. (**a**) The photodetector output signal is digitized and recorded by an ADC. This signal shows sequential laser pulses. (**b**) Each pulse is saved separately as a frame for further processing. (**c**) The analytic form of high-frequency components of each pulse is generated using Hilbert transformation, and the phase component of this analytic form is extracted. (**d**) An unwrapping algorithm is used to fix unrealistic phase jumps, and the result shows an approximately linear phase increase. (**e**) If the phase component of the interferometer fringe frequency is removed, the phase induced by cells in optical pulse can be seen. (**f**) Many of these line images generated from subsequent frames are used to form a spatial map of optical path difference in two dimensions, which is used for cell characterization

reach the photodetector serially. An analog-to-digital converter (ADC) with a sampling rate of 50 GSps and 20 GHz bandwidth is used to acquire the output signal of the photodetector.

The photodetector output signal, $I(t)$, is digitized and recorded by the ADC (Fig. 5.2a). This signal shows sequential laser pulses. Each pulse is used to form one line image. Therefore, the boundaries of pulses are determined precisely, and each pulse is saved separately as a frame for further processing (Fig. 5.2b). The analytic form of each pulse is generated using Hilbert transformation after the low frequency components corresponding to intensity variations are filtered out [92]. The phase component of this analytic form is extracted, while its amplitude component is discarded (Fig. 5.2c). Because the phase varies over a wide range (much larger than 2π radians), it shows unrealistic discontinuities. An unwrapping algorithm is used to fix these discontinuities, and the result shows an approximately linear phase increase over the time for each pulse or frame (Fig. 5.2d). The unwrapping algorithm adds multiples of $\pm 2\pi$ to make the absolute jumps between consecutive samples in a frame smaller than π radians when they are greater than π radians. If the linear component of the phase, which corresponds to the fringe (modulation) frequency, f_m, due to the interferometer arms' length mismatch, and the background phase level,

φ_0, are subtracted, the phase shift induced by the cells in the optical pulse can be observed (Fig. 5.2e); i.e.

$$\Delta\varphi(t) = \text{unwrap}(\arg(I_{\text{BP}}(t) + j \cdot \hat{I}_{\text{BP}}(t))) - 2\pi f_m t - \varphi_0 \tag{5.1}$$

in which $I_{\text{BP}}(t)$ is a band-pass filtered form of $I(t)$ with only spectral features modulated at f_m, and $\hat{I}_{\text{BP}}(t)$ is the Hilbert transform of $I_{\text{BP}}(t)$. Many phase line images generated from subsequent frames are combined to form a spatial map of optical path difference (OPD) in two dimensions (Fig. 5.2f). Since we know the mapping of space to time from the rainbow characteristics and flow speed, OPD at each point is calculated as

$$\text{OPD}(x, y) = \frac{\lambda(x)}{2\pi}\Delta\varphi(x, y) \tag{5.2}$$

where x and y are coordinates in the rainbow and flow directions, respectively; $\lambda(x)$ is the wavelength at position x along the rainbow; and $\Delta\varphi(x, y)$ is the phase shift induced by the cell at point (x, y).

Spatial map of optical path difference can be used to extract the refractive index contrast between the cell and the surrounding liquid. If the thickness of the cell at point (x, y) is $t(x, y)$,

$$\text{OPD}(x, y) = 2\Delta n_{\text{cell}} \cdot t(x, y) \tag{5.3}$$

where $\Delta n_{\text{cell}} = n_{\text{cell}} - n_{\text{liquid}}$ in which n_{cell} and n_{liquid} are the refractive indices of the cell and the surrounding liquid, respectively. The factor 2 is to account for the fact that each wavelength component passes the cell twice in Michelson interferometer. If we integrate Eq. (5.3) over the area of the cell, we can derive an average refractive index contrast, which corresponds to protein concentration of the cell:

$$\Delta n_{\text{cell}} = \frac{\iint_{\text{cell}} \text{OPD}(x, y)\, \mathrm{d}x\, \mathrm{d}y}{2V_{\text{cell}}} \tag{5.4}$$

where $V_{\text{cell}} = \iint_{\text{cell}} t(x, y)\, \mathrm{d}x\, \mathrm{d}y$ is the volume of the cell. Most of the cells relax to a spherical shape when they are released from substrates and brought into suspension [93, 94]. Therefore, if we know the diameter of the cell, d_{cell}, we can estimate its volume as $V_{\text{cell}} \approx \pi d_{\text{cell}}^3/6$.

5.3 Results and Discussion

Spherical polystyrene beads with a NIST traceable diameter of 5 μm are used to calibrate the image processing algorithm for size measurements. A custom designed algorithm in CellProfiler software [33] is used to detect the beads or cells in spatial

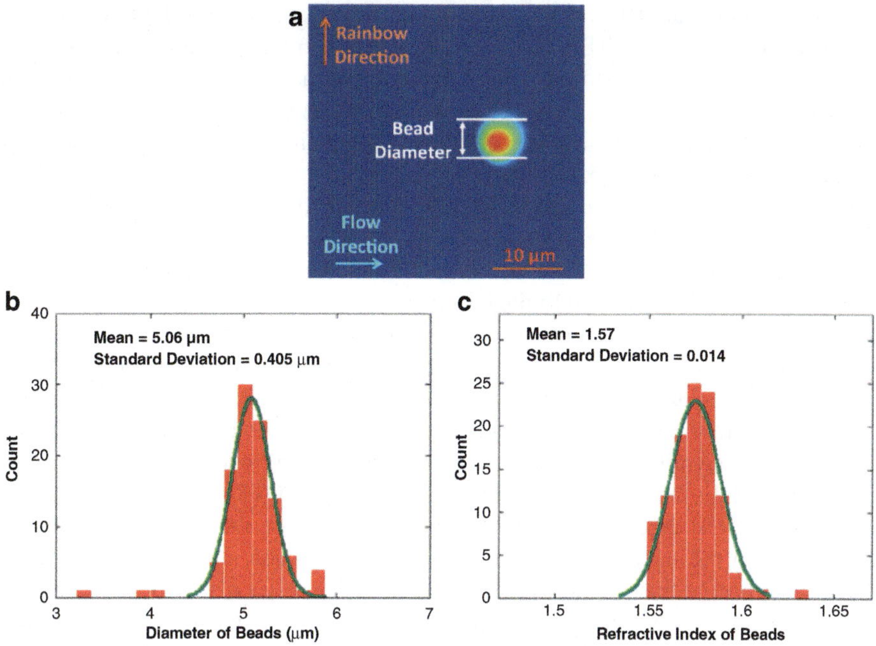

Fig. 5.3 Calibration with NIST traceable beads. Polystyrene beads with a NIST traceable diameter of 5 μm are used to calibrate the image processing algorithm for size measurements. (**a**) A custom designed image processing algorithm in CellProfiler software is used to find the beads in spatial map of optical path difference and measure the diameter. (**b**) Histogram of bead diameters demonstrates the measured size distribution has an expected mean of 5 μm and a standard deviation within the range of optical resolution limit. (**c**) Since all the beads are made out of the same material, the coefficient of variation for refractive indices ($0.014/1.57 = 0.89\%$) is much smaller than that of diameters ($0.405/5.06 = 8.00\%$)

map of optical path difference (Fig. 5.3). Bead or cell diameter is measured along the rainbow direction to eliminate size measurement inaccuracies caused by fluctuations of flow speed (Fig. 5.3a). Due to limited optical resolution of the setup, the bead or cell edges are blurred, generating a small phase signal outside of the diameter bars. The diameter along the rainbow direction is equal to the diameter along the interrogation optical beam for spherical-shape beads or cells in suspension, including the samples in our experiments.

Histogram analysis of bead diameter distribution for more than one hundred beads with corresponding Gaussian fit to measurements demonstrates that the measured size distribution has a standard deviation of 0.4 μm and an expected mean of 5 μm (Fig. 5.3b). The broadening in the distribution is caused by the limited lateral optical resolution of the Coherent-STEAM setup. This resolution is measured by the knife-edge method and is about 2.5 μm. Therefore, the standard deviation of the bead size distribution is well below the optical resolution.

We also measured the refractive index contrast of each bead and the surrounding liquid using Coherent-STEAM. Assuming that the refractive index of water is 1.317 at the 1581–1601 nm bandwidth, we derived the refractive index of the beads using Eq. (5.4). Analysis of the bead refractive indices and corresponding Gaussian fit demonstrates that the beads have a mean refractive index of 1.57 with a standard deviation of 0.014 (Fig. 5.3c). We observe that the coefficient of variation for the bead refractive indices is 0.89%, which is much smaller than the coefficient of variation for the bead diameters (8.00%). This is expected because all the beads are made out of the same material, while their diameter measurements are effected by dispersity of the size and limited spatial resolution of the setup.

We used the calibrated Coherent-STEAM setup to measure cell diameter and refractive index contrast (as a measure for protein concentration) simultaneously. Different types of cells have different mean diameters and protein concentrations; however, both of these parameters have a broad range of variations for each cell type. We see that identification of cells is more specific using both of these parameters simultaneously, instead of each individually. Images of OTII (Fig. 5.4a) and SW480 (Fig. 5.4b) cells taken by Coherent STEAM setup demonstrate that the cells are spherical in the microfluidic channel. In Fig. 5.4c, scattering plot of cell protein concentration (refractive index difference) versus diameter is shown for these cells. Using points in a normal range of protein concentration and sliding the detection

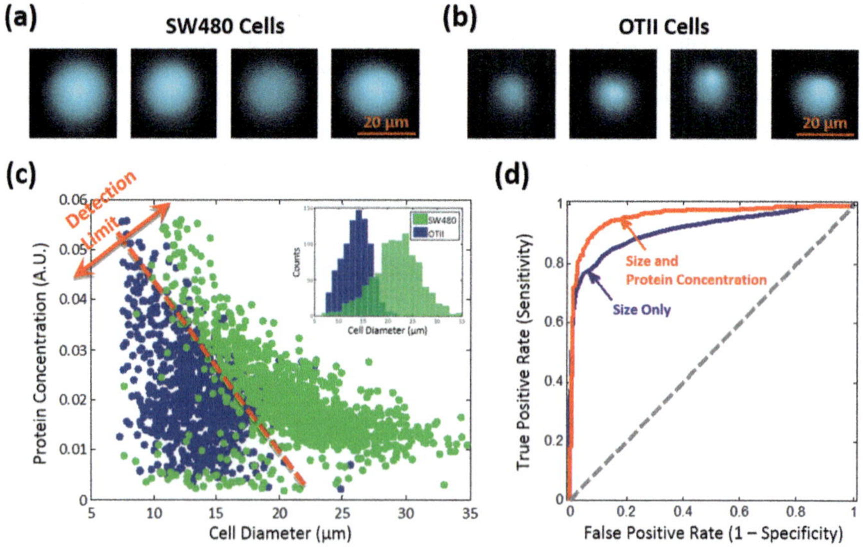

Fig. 5.4 Cell classification based on size and protein concentration measurement by Coherent-STEAM; images of (**a**) SW480 and (**b**) OTII cells taken by Coherent STEAM setup show that they are *spherical*. (**c**) Scattering plot of cell protein concentration versus diameter is shown for OTII (*blue*) and SW480 (*green*) cells. (**d**) Comparison of the ROC curves of size measurement only (*purple line*) to that of simultaneous size and protein concentration measurement (*orange line*) shows significant improvement in sensitivity

limit along the depicted direction (perpendicular to the optimum classification line), a receiver operating characteristic (ROC) curve is generated (Fig. 5.4d). Comparing the ROC curve of individual parameters (e.g., size measurement only) to that of simultaneous measurement, it becomes obvious that the detection sensitivity has improved considerably.

5.4 Conclusion

In summary, we demonstrated a new type of imaging flow cytometry based on coherent stretched-time-encoded amplified microscopy, which is capable of classifying cells in flow rates as high as a few meters per second. Coherent-STEAM measures size and total optical path difference of cells simultaneously and extracts the refractive index, which corresponds to the protein concentration of the cells, as an additional parameter for classification. As illustrated in our experimental results, separation of two cell types was significantly enhanced by adopting the additional protein concentration parameter generated by Coherent-STEAM. We will continue our work with real-time signal processing and cell identification on field-programmable gate arrays (FPGAs) for classification of more than two cell types.

Chapter 6
Time Stretch Quantitative Phase Imaging

Label-free cell analysis is essential to personalized genomics, cancer diagnostics, and drug development as it avoids adverse effects of staining reagents on cellular viability and cell signaling. However, currently available label-free cell assays mostly rely only on a single feature and lack sufficient differentiation. Also, the sample size analyzed by these assays is limited due to their low throughput. Here, we integrate feature extraction and deep learning with high-throughput quantitative imaging enabled by photonic time stretch, achieving record high accuracy in label-free cell classification. Our system captures quantitative optical phase and intensity images and extracts multiple biophysical features of individual cells. These biophysical measurements form a hyperdimensional feature space in which supervised learning is performed for cell classification.

6.1 Background

Flow cytometry is a powerful tool for large-scale cell analysis due to its ability to measure anisotropic elastic light scattering of millions of individual cells as well as emission of fluorescent labels conjugated to cells [95, 96]. However, each cell is represented with single values per detection channels (forward scatter, side scatter, and emission bands) and often requires labeling with specific biomarkers for acceptable classification accuracy [95, 97]. Imaging flow cytometry [98, 99] on the other hand captures images of cells revealing significantly more information about the cells. For example, it can distinguish clusters and debris that would otherwise result in false positive identification in a conventional flow cytometer based on light scattering [33].

In addition to classification accuracy, another critical specification of a flow cytometer is its throughput. Indeed high throughput, typically 100,000 cells per second, is needed to screen a large enough cell population to find rare abnormal

A. Mahjoubfar et al., *Artificial Intelligence in Label-free Microscopy*,
DOI 10.1007/978-3-319-51448-2_6

cells that are indicative of early stage diseases. However there is a fundamental trade-off between throughput and accuracy in any measurement system [29, 100]. Additionally, imaging flow cytometers face a throughput limit imposed by the speed of the CCD or the CMOS cameras, a number that is approximately 2000 cells/s for present systems [30]. Higher flow rates lead to blurred cell images due to the finite camera shutter speed. Many applications of flow analyzers such as cancer diagnostics, drug discovery, biofuel development, and emulsion characterization require classification of large sample sizes with a high-degree of statistical accuracy [101]. This has fueled research into alternative optical diagnostic techniques for characterization of cells and particles in flow.

In previous chapter, a label-free imaging flow-cytometry technique based on coherent optical implementation of the photonic time stretch concept [17] has been presented. This instrument overcomes the trade-off between sensitivity and speed by using Amplified Time stretch Dispersive Fourier Transform [9, 47, 102]. In time stretched imaging [10], the object's spatial information is encoded in a spectrum of laser pulses within a pulse duration of sub-nanoseconds (Fig. 6.1). Each pulse representing one frame of the camera is then stretched in time so that it can be digitized in real-time by an electronic analog-to-digital converter (ADC). The ultra-fast pulse illumination freezes the motion of high-speed cells or particles in flow to achieve blur-free imaging. Detection sensitivity is challenged by the low number of photons collected during the ultra-short shutter time (optical pulse width) and the drop in the peak optical power resulting from the time stretch. These issues

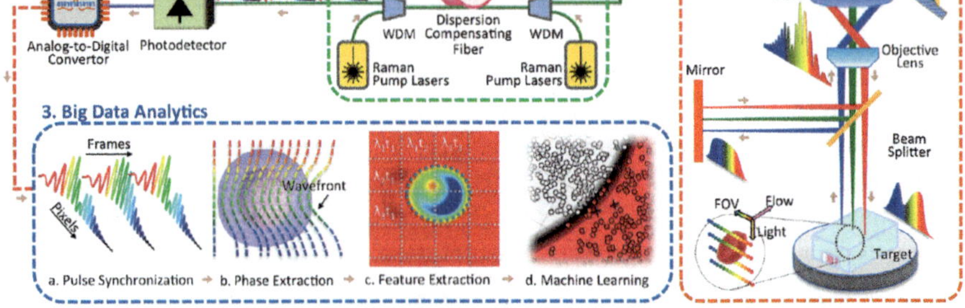

Fig. 6.1 Time stretch quantitative phase imaging (TS-QPI) and analytics system. A mode-locked laser followed by a nonlinear fiber, an erbium doped fiber amplifier (EDFA), and a wavelength-division multiplexing (WDM) filter generate and shape a train of broadband optical pulses. *Box 1*: The phase shift and intensity loss at each location within the field of view are embedded into the spectral interference patterns using a Michelson interferometer. *Box 2*: Time stretch dispersive Fourier transform. *Box 3*: Big data analytics pipeline

are solved in time stretch imaging by implementing a low noise-figure Raman amplifier within the dispersive device that performs time stretching [10, 17, 29]. Moreover, warped stretch transform [103] can be used in time stretch imaging to achieve optical image compression and nonuniform spatial resolution over the field-of-view [104]. In the coherent version of the instrument, the time stretch imaging is combined with spectral interferometry to measure quantitative phase and intensity images in real-time and at high throughput [19]. Integrated with a microfluidic channel, coherent time stretch imaging system in this work measures both quantitative optical phase shift and loss of individual cells as a high-speed imaging flow cytometer, capturing 36 million images per second in flow rates as high as 10 m/s, reaching up to 100,000 cells per second throughput.

On another note, surface markers used to label cells, such as EpCAM [105], are indispensable tools for cell classification. However, they are unavailable in some applications; for example, melanoma or pancreatic circulating tumor cells (CTCs) as well as some cancer stem cells are EpCAM-negative and will escape EpCAM-based detection platforms [31]. Furthermore, large-population cell sorting opens the doors to downstream operations, where the negative impacts of labels on cellular behavior and viability are often unacceptable [106]. Cell labels may cause activating/inhibitory signal transduction, altering the behavior of the desired cellular subtypes, potentially leading to errors in downstream analysis, such as DNA sequencing and subpopulation regrowth. In this way, quantitative phase imaging (QPI) methods [92, 107, 108] that categorize unlabeled living cells with high accuracy are needed. Coherent time stretch imaging is a method that enables quantitative phase imaging at ultrahigh throughput for noninvasive label-free screening of large number of cells.

In this work, the information of quantitative optical loss and phase images are fused into expert designed features, leading to a record label-free classification accuracy when combined with deep learning (details about deep learning can be found in next chapter). Image mining techniques are applied, for the first time, to time stretch quantitative phase imaging to measure biophysical attributes including protein concentration, optical loss, and morphological features of single cells at an ultrahigh flow rate and in a label-free fashion. These attributes differ widely [87, 109–111] among cells and their variations reflect important information of genotypes and physiological stimuli [112].

6.2 Time Stretch Quantitative Phase Imaging

6.2.1 Overview

The application of time stretch quantitative phase imaging (TS-QPI) to imaging flow cytometry has been recently demonstrated in our group [17]. Broadband optical pulses from a mode-locked laser were firstly conditioned in fiber optics and then spatially dispersed in free-space optics with a pair of reflection diffraction gratings creating 1-D "rainbow flashes" (Fig. 6.1). Each one of rainbow flashes was composed of all the wavelength components distributed laterally over the field of view. These flashes illuminated the target as in traditional photography, but in addition, rainbow flashes targeted different spatial points with distinct colors of light, resulting in space-to-spectrum encoding. Rainbow pulses were then split into the two arms of a Michelson interferometer. Different wavelength components of the rainbow flash traveled parallel to each other but individually focused on the mirror in the reference arm or on the reflective substrate of a microfluidic device in the sample arm. In the sample arm, the cells in the microfluidic channel were hydrodynamically focused [113, 114] into the rainbow's field of view and flowed perpendicular to the rainbow flash. Reflected pulses from the microfluidic device and the reference arm were recombined and coupled back into the fiber, optically amplified and linearly chirped through Raman-amplified time stretch dispersive Fourier transform (TS-DFT) system. An amplified time stretch system that utilizes a low-noise distributed Raman amplifier within dispersive fiber with a net optical gain of approximately 15 dB enables high-sensitivity detection at high speeds. An ultrafast single-pixel photodetector transformed instantaneous optical power into an electrical signal and subsequently, an analog-to-digital converter (ADC) samples and quantizes the signal. Acquired data are passed down to processing stages for big data analytics. The interference between time-shifted linearly chirped pulses creates a beat (fringe) frequency, which can be adjusted via the interferometer arm length mismatch.

In Fig. 6.1 Box 1, the pulse train is spatially dispersed into a train of rainbow flashes illuminating the target as line scans. The spatial features of the target are encoded into the spectrum of the broadband optical pulses, each representing a one-dimensional frame. The ultra-short optical pulse illumination freezes the motion of cells during high-speed flow to achieve blur-free imaging with a throughput of 100,000 cells/s. The phase shift and intensity loss at each location within the field of view are embedded into the spectral interference patterns using a Michelson interferometer. In Box 2, the interferogram pulses were then stretched in time so that spatial information could be mapped into time through time stretch dispersive Fourier transform (TS-DFT), and then captured by a single pixel photodetector and an analog-to-digital converter (ADC). The loss of sensitivity at high shutter speed is compensated by stimulated Raman amplification during time stretch. In Box 3a, the time-domain signal carrying serially captured rainbow pulses is transformed into a series of one-dimensional spatial maps, which are used for forming line images. The biomass density of a cell leads to a spatially varying optical phase shift (Box 3b). When a rainbow flash passes through the cells, the changes in refractive index at different locations will cause phase walk-off at interrogation wavelengths. Hilbert transformation and phase unwrapping are used to extract the spatial phase shift. Decoding the phase shift in each pulse at each wavelength and remapping it into a pixel reveal the protein concentration distribution within cells (Box 3c). The optical loss induced by the cells, embedded in the pulse intensity variations, is obtained from the amplitude of the slowly varying envelope of the spectral interferograms. Thus, quantitative optical phase shift and intensity loss images are captured simultaneously. Both images are calibrated based on the regions where the cells are absent. Cell features describing morphology, granularity, biomass, etc. are extracted from the images. These biophysical features are used in a machine learning algorithm for high-accuracy label-free classification of the cells (Box 3d).

The photodetected time stretched pulses, each representing one line scan, are converted to analytic signals using Hilbert transformation [115] and the intensity and phase components are extracted. The phase component is a fast oscillating fringe (carrier frequency), caused by the interference of the linearly chirped pulses arriving from the reference and signal arms in the Michelson interferometer. Acting as a radio-frequency (RF) carrier whose frequency is set by the user adjusted arm length mismatch, the fringe frequency is modulated when the optical path length in the sample arm is changed by the arrival of a cell. This frequency shift and the accompanying phase change are used to measure the optical path length of the cell. Since the phase varies over a wide range (much larger than 2π radians), an unwrapping algorithm is used to obtain the continuous phase profile. The phase profile contains an increasing term with time, corresponding to the fringe (beat) frequency and the phase shift induced by the cell. By eliminating the background phase component, the cell-induced phase shift is extracted. The second component in the waveform is a lower frequency envelope corresponding to temporal shape of the optical pulse. The amplitude of this envelope provides information about optical loss caused by transparency, surface roughness, and inner organelle complexity.

6.2.2 Imaging System

Broadband optical pulses from a mode-locked laser (center wavelength = 1565 nm, repetition rate = 36.6 MHz, pulse width $\approx$ 100 fs) are broadened using a highly nonlinear fiber to approximately 100 nm bandwidth with a spectral range up to 1605 nm. These broadband pulses are then linearly chirped to nanosecond pulse width by a short dispersion compensating fiber (DCF) of 60 ps/nm, so that an erbium doped fiber amplifier (EDFA) can amplify them with minimal distortion. A coarse wavelength division multiplexer (WDM) filters the pulses from 1581 nm to 1601 nm, where the spectrum is reasonably flat. Therefore, the total bandwidth of the pulses interrogating the cells in our setup is less than 20 nm centered at 1591 nm, giving a negligible fractional bandwidth of 1.3%. These filtered pulses then pass through an optical circulator and are coupled to free-space with a fiber collimator.

Free-space laser pulses were linearly polarized with quarter- and half-wave plates, and then spatially dispersed with a pair of reflection diffraction gratings, so that each wavelength component of the collimated beam was positioned at a different lateral point similar to a line flash rainbow. A beam reducer shrank the rainbow beam six times with a pair of 90 degree off-axis parabolic gold-coated mirrors with reflected focal lengths as 152.4 mm and 25.4 mm, respectively. Next, a 15 degree off-axis parabolic gold-coated mirror with 635 mm reflected focal length and a long working-distance objective lens with 0.4 numerical aperture further

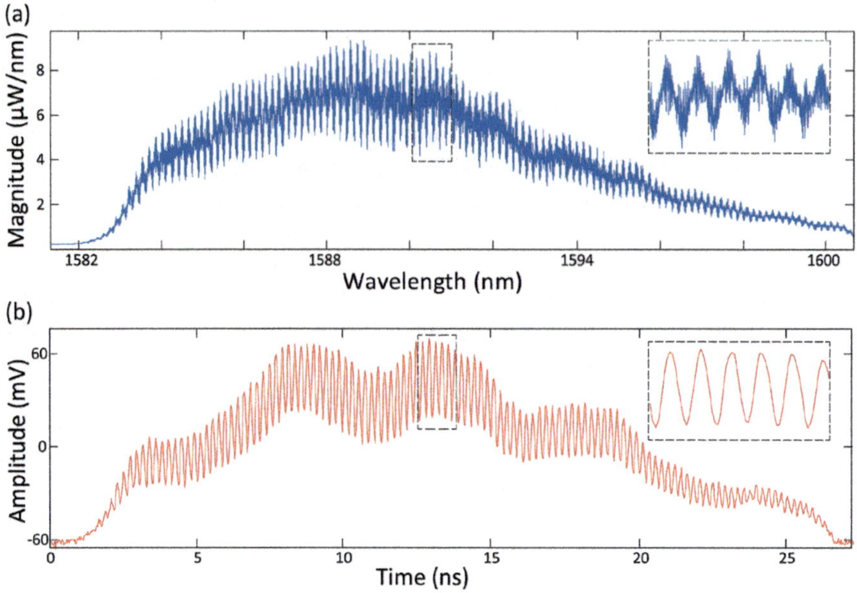

Fig. 6.2 Comparison of the interferograms measured by optical spectrum analyzer and time stretch dispersive Fourier Transform; (**a**) Optical spectrum of the signal after quantitative phase imaging (box 1 in Fig. 6.1) and before it enters the amplified time stretch system (box 2 in Fig. 6.1). The interference pattern in spectral domain is measured by an optical spectrum analyzer. (**b**) With time stretch, the interference pattern in spectral domain is linearly mapped into time. The baseband intensity envelope is slightly modified by the wavelength-dependent gain profile of the Raman amplifier. The inserts in panels **a** and **b** show the zoomed-in spectrum and waveform in the *dashed black boxes*, respectively. Clearly, the single-shot interferogram measured by Raman-amplified time stretch dispersive Fourier Transform has a higher signal-to-noise ratio compared to that captured by optical spectrum analyzer

shrank the rainbow to about 130 μm in width, i.e., field of view. Reflective optics with parabolic gold-coated mirrors is used in our experimental demonstration to minimize loss, aberration, and polarization sensitivity. The rainbow flashes were then split into the two arms of a Michelson interferometer by a beam splitter. In the sample arm, the rainbow pulses pass through the cells and are reflected by the reflective substrate of the microfluidic device. In the reference arm, a dielectric mirror reflected the rainbow with a length mismatch with the sample arm causing spectral interference fringes (Fig. 6.2a). Cells are hydrodynamically focused at the center of the channel flow at a velocity of 1.3 m/s. The reflected pulses from reference and sample arms were recombined at the beam splitter, compressed by the two diffraction gratings and coupled back into the fiber. These return pulses were spectrally encoded by the spatial information of the interrogation field of view. Then they were redirected by the optical circulator to a Raman-amplified time stretch dispersive Fourier Transform (TS-DFT) system followed by a 10 Gb/s photodetector (Discovery Semiconductors DSC-402APD). An analog-to-digital converter (Tektronix DPO72004C) with a sampling rate of 50 GS/s and 20 GHz bandwidth is used to acquire the output signal of the photodetector, which is a series of spectral interferograms mapped into time (Fig. 6.2b).

6.2.3 *System Performance and Resolvable Points*

Lateral resolution of time stretch camera is decided by the limiting factor among Abbe diffraction limit of the objective lens, spectral resolvability of the diffraction grating pairs, spectral resolution in amplified dispersive Fourier transform, the photodetector rise-time and bandwidth, and the sampling rate of the back-end digitizer. Details of the limiting factors of lateral resolution and evaluation of these factors for our TS-QPI system can be found in Table 6.1. Field of view (FOV) is the area covered by the interrogation rainbow when the rainbow pulses hit the imaging plane. The rainbow pulse width is decided by the optical bandwidth selected from the laser source, $\Delta\lambda$, the magnification factor of the objective lens, the focal length of the other lenses and parabolic mirrors, as well as the dimensions and blaze angles of the diffraction gratings.

The resolution of phase measurement along axial direction is determined by the effective number of bits (ENOB) of the digitizer and affected by the noise of laser source. Since pulse-to-pulse intensity and phase fluctuations are small, noise from laser source is not the limiting factor in our phase measurements. Supposing the ENOB of the digitizer is N, the minimum detectable optical path length difference ΔL can be estimated as

$$\frac{1}{2}\sin\left(\frac{4\pi\Delta L}{\lambda + \Delta\lambda/2}\right) = 2^{-N} \tag{6.6}$$

where λ is the central wavelength of light, and $\Delta\lambda$ is the optical bandwidth. In our system, ENOB of the analog-to-digital converter is 5. Thus, the OPD resolution along the axial direction is about 8.0 nm, corresponding to refractive index difference down to the order of 0.001 for cellular level measurements.

6.2.4 *Microfluidic Channel Design and Fabrication*

The Polydimethylsiloxane (PDMS) microfluidic channel is custom-designed so that it could fit into the reflective optics design. Cells are hydrodynamically focused [113, 114] at the center of the channel flowing at a velocity of 1.3 m/s. The microfluidic device consists of a hydrodynamic focusing region and an imaging region targeted by the interrogation rainbow flashes in TS-QPI system. At the hydrodynamic focusing region, the sheath pressure focused the sample at the center

Table 6.1 Resolution limiting factors in TS-QPI

System category	Component	Number of resolvable points	Lateral resolution
Free-space optics	Diffraction gratings	$$N_{\text{grating}} = \frac{\Delta\lambda}{\delta\lambda_{\text{grating}}} = \Delta\lambda / \left(\lambda \cdot \frac{d}{m \cdot 2w_0}\right) \quad (6.1)$$ where $\Delta\lambda$ is the optical bandwidth, λ is the central wavelength, m is the order of diffraction, w_0 is the beam waist, and d is the groove spacing	3.09 μm
	Lenses and mirrors	$$N_{\text{Abbe}} = \frac{\text{FOV}}{\delta x_{\text{diffraction}}} = \frac{\text{FOV}}{\left(\frac{\lambda+\Delta\lambda/2}{2\cdot\text{NA}}\right)} \quad (6.2)$$ where FOV is field of view, NA is numerical aperture of the objective lens	2.00 μm
Time stretch	Group delay dispersion	$$N_{\text{DFT}} = \frac{\Delta\lambda}{\delta\lambda} = \frac{\Delta\lambda}{\lambda \cdot \sqrt{\frac{2}{DL_f \cdot c}}} \quad (6.3)$$ where D is the group velocity dispersion, L_f is the dispersive fiber length	0.73 μm
Electronic back-end	Photodetector bandwidth	$$N_{\text{PD}} = \frac{\Delta t}{\delta t} = \frac{DL_f\Delta\lambda}{0.35/B} \quad (6.4)$$ where B is the bandwidth of the photodetector	0.28 μm
	ADC sampling rate	$$N_{\text{ADC}} = DL_f\Delta\lambda f_{\text{ADC}} \quad (6.5)$$ where f_{ADC} is the sampling rate of digitizer	0.10 μm

of the channel by narrowing its flow width from 200 μm to about 40 μm with a sheath to sample volume ratio of 3:1. The dimension of the channel was chosen as 200 μm (width) × 25 μm (height) so that the cells will be imaged within depth of focus with a narrow lateral distribution. The size of the entire PDMS channel is optimized for fitting on a 2-inch diameter dielectric mirror with sufficient space at the edges to achieve strong bonding. The thickness of the channel top layer is optimized for stabilizing peek tubes performance reliability while accommodating the working distance of the objective lens.

The PDMS microfluidic channel (Fig. 6.3) is fabricated using standard soft lithography. The mask was designed in AutoCAD and printed with a resolution down to 1 μm. Then a 4-inch silicon wafer was spin-coated with 75 μm thickness of a negative photoresist (SU-8 from MicroChem) and was exposed under the mask using an aligner. After post-exposure baking, the wafer was developed at room temperature, rinsed with isopropyl alcohol (IPA), and placed in a petri dish. A PDMS mixture (Sylgard 184 Silicone Elastomer, Dow Corning) was poured onto the patterned wafer, degassed in a vacuum chamber for 30 min, and cured at 80 °C

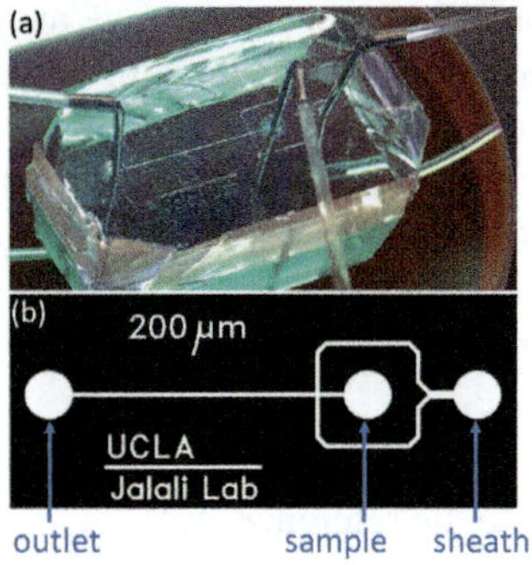

Fig. 6.3 PDMS microfluidic channel mounted on a highly reflective surface with near-infrared dielectric coating; The microfluidic device consists of a hydrodynamic focusing region and an imaging region targeted by the interrogation rainbow flashes in TS-QPI system. (**a**) Sample solution with suspended cells is fed into the channel through the sample inlet, and deionized water as the sheath fluid is injected through the sheath inlet. At the hydrodynamic focusing region, the sheath pressure focused the sample at the center of the channel by narrowing its flow width from 200 μm to about 40 μm with a sheath to sample volume ratio of 3:1. (**b**) The pattern of the mask used to imprint microfluidic channel design on silicon wafer with photoresist. The *circles* are inlet and outlet reservoirs

for 1 h. Once cured, the PDMS channel was cut out and peeled off from the master wafer. We used 1.25 μm diameter hollow needle to punch the inlet and outlet holes. The punched PDMS channel was then cleaned with nitrogen gun and magic tape (3M), treated with oxygen plasma (Enercon Dyne-A-Mite 3D Treater) for 2 min, and bonded to a 2-inch diameter broadband dielectric mirror (Thorlabs BB2-E04) for obtaining high reflectance from channel substrate at near infrared spectral window. Finally microtubes (PE-50 tubing, 0.023 × 0.038 in) with steel catheter couplers (Instech, 22 ga ×15 mm) are connected to the inlet and outlet punctures.

6.2.5 *Coherent Detection and Phase Extraction*

Unlike in conventional heterodyne detection, which uses a narrowband continuous-wave signal as the local oscillator or reference, the coherent detection in our time stretch system uses an unmodulated copy of the original optical input, which is a broadband optical pulse train [116, 117].

Since the spectrum is mapped into space by diffraction gratings, the complex field at any specific spatial location within the field of view is a narrowband optical wave. As the envelope of the optical wave varies slowly in time compared to the period of the optical electromagnetic wave and the time mismatch between the reference arm and the sample arm, we employ slowly varying envelope approximation in our analysis. The complex envelope of the input electric field, $\tilde{E}_{in}(\omega, t_p)$, is split into two arms of the Michelson interferometer at the beam splitter. Here, ω is the optical frequency of the input signal, which corresponds to the spatial location x being interrogated by the optical wave at this frequency (i.e., spectral encoding of the object image). t_p specifies the time when each rainbow flash reaches the beam splitter, corresponding to the p-th incoming pulse. Note that $\tilde{E}_{in}(\omega, t_p)$ can be simplified as $\tilde{E}_{in}(\omega)$ when pulse shape is stable from pulse to pulse. The light split into the two arms of the Michelson interferometer can be expressed as

$$\begin{aligned}
&\text{Into the sample arm:}\\
&\quad \tilde{E}_s(\omega, t_p) = \sqrt{T_b}\tilde{E}_{in}(\omega, t_p)\\
&\text{Into the reference arm:}\\
&\quad \tilde{E}_r(\omega, t_p) = i\sqrt{1 - T_b}\tilde{E}_{in}(\omega, t_p)
\end{aligned} \tag{6.7}$$

where T_b is the power transmission ratio of the beam-splitter. Optical intensity in the sample arm will be altered by the absorption and scattering of imaged cells, as well as that of the microfluidic channel and buffer solution. Not only the electric field amplitude after passing through semitransparent objects will be modulated by the optical attenuation in the sample arm, but also the optical path length difference will lead to a phase shift, $\Delta\varphi_c(x, t_p)$, induced by refractive index change from the object along the interrogation beam. Thus, the complex fields of the light waves coming back to the beam splitter become

$$\begin{aligned}
&\text{From the sample arm:}\\
&\quad \tilde{E}_s(\omega, t_p) = T_s(\omega)T_c(x, t_p + t_d)\sqrt{T_b(\omega)R_m(\omega)}\tilde{E}_{in}(\omega, t_p)\\
&\qquad\qquad \cdot \exp\left\{i\left[-\frac{\omega}{c}\cdot 2(L + \Delta L) - \Delta\varphi_c(x, t_p)\right]\right\}\\
&\text{From the reference arm:}\\
&\quad \tilde{E}_r(\omega, t_p) = i\sqrt{1 - T_b(\omega)}\sqrt{R_m(\omega)}\tilde{E}_{in}(\omega, t_p)\\
&\qquad\qquad \cdot \exp\left(-i\frac{\omega}{c}\cdot 2L\right)
\end{aligned} \tag{6.8}$$

where L is the length of reference arm, and ΔL is the arm length mismatch between two arms. $R_m(\omega)$ is the wavelength-dependent reflectance of the reflective substrate of the microfluidic channel and the dielectric mirror in the reference arm. t_d is the time delay during which rainbow flash travels from the beam splitter to the sample cell, $t_d = \frac{L_0+\Delta L}{c}$. $T_s(\omega)$ is power transmittance of the surrounding buffer solution and microfluidic channel, and $T_c(x, t_p + t_d)$ is spatial power transmittance of cells at location x along the rainbow when being illuminated at time $t_p + t_d$. Both $\sqrt{T_s(\omega)}$ and $\sqrt{T_c(x, t_p + t_d)}$ affect the optical field twice as each rainbow flash passes through the cell twice. Since the t_d is much smaller than the time scale of the envelope variations caused by the cell flow, we can approximate $T_c(x, t_p + t_d)$ to be $T_c(x, t_p)$ to synchronize with $\tilde{E}_{in}(\omega, t_p)$ without sacrificing accuracy.

The total electric field at each wavelength or optical frequency after two arms of the interferometer recombine at the beam splitter becomes

$$
\begin{aligned}
\tilde{E}_o(\omega, t_p) &= T_s(\omega)T_c(x, t_p)T_b(\omega)\sqrt{R_m(\omega)}\tilde{E}_{in}(\omega, t_p) \\
&\quad \cdot \exp\left\{i\left[-\frac{\omega}{c}\cdot 2(L+\Delta L) - \Delta\varphi_c(x, t_p)\right]\right\} \\
&\quad - [1 - T_b(\omega)]\sqrt{R_m(\omega)}\tilde{E}_{in}(\omega, t_p)\exp\left[-i\frac{\omega}{c}\cdot 2L\right] \\
&= \sqrt{R_m(\omega)}\tilde{E}_{in}(\omega, t_p)\exp\left(-i\frac{2\omega L}{c}\right) \\
&\quad \cdot \left\{T_s(\omega)T_c(x, t_p)T_b(\omega)\exp\left\{i\left[-2\frac{\omega}{c}\Delta L\right.\right.\right. \\
&\quad \left.\left.\left. -\Delta\varphi_c(x, t_p)\right]\right\} - [1 - T_b(\omega)]\right\}.
\end{aligned}
\tag{6.9}
$$

Based on the spectral encoding setup, we know spatial information has been encoded into spectrum,

$$T_c(x, t_p) \Rightarrow T_c(\omega, t_p) \tag{6.10}$$

$$\Delta\varphi_c(x, t_p) \Rightarrow \Delta\varphi_c(\omega, t_p) \tag{6.11}$$

The intensity envelope then becomes

$$
\begin{aligned}
\tilde{I}_o(\omega, t_p) &\propto \tilde{E}_o^*(\omega, t_p)\tilde{E}_o(\omega, t_p) \\
&= \left\{[1 - T_b(\omega)]^2 + T_b^2(\omega)T_c^2(\omega, t_p)T_s^2(\omega)\right\}R_m(\omega) \\
&\quad \cdot |\tilde{E}_{in}(\omega, t_p)|^2 - 2[1 - T_b(\omega)]T_c(\omega, t_p)T_b(\omega)R_m(\omega) \\
&\quad \cdot T_s(\omega)|\tilde{E}_{in}(\omega, t_p)|^2\cos[2\omega\Delta L/c + \Delta\varphi_c(\omega, t_p)]
\end{aligned}
\tag{6.12}
$$

During time stretch, each frequency component ω or wavelength λ will be one-to-one mapped into time domain. We define the relative time delay of λ compared to the central wavelength, λ_c, as t_i, which is usually called intra-pulse time delay. Written in terms of λ, Eq. (6.12) can be simplified as

$$\tilde{I}_o(\lambda, t_p) \propto I_b(\lambda, t_p) + I_i(\lambda, t_p)\cos\left[\frac{4\pi\Delta L}{\lambda} + \Delta\varphi_c(\lambda, t_p)\right] \tag{6.13}$$

where $I_b(\lambda, t_p)$ is the background or baseband intensity envelope, and $I_i(\lambda, t_p)$ is the interference or intermediate intensity envelope:

$$I_b(\lambda, t_p) = \left\{[1 - T_b(\lambda)]^2 + T_b^2(\lambda)T_c^2(\lambda, t_p)T_s^2(\lambda)\right\} \cdot R_m(\lambda)|\tilde{E}_{in}(\lambda, t_p)|^2 \tag{6.14}$$

$$I_i(\lambda, t_p) = -2[1 - T_b(\lambda)]T_c(\lambda, t_p)T_b(\lambda)R_m(\lambda) \cdot T_s(\lambda)|\tilde{E}_{in}(\lambda, t_p)|^2 \tag{6.15}$$

Linear time stretch maps frequency domain into time domain by

$$t_i = D(\lambda - \lambda_c)L_f. \tag{6.16}$$

where λ_c is the central wavelength and L_f the length of the dispersive fiber. D is the group velocity dispersion, that is, the temporal pulse spreading, Δt_i, per unit bandwidth, $\Delta\lambda$, per unit distance traveled. Thus the temporal samples of the energy flux absorbed at the photodetector are the intra-pulse concatenation of spectral samples followed by inter-pulse concatenation of pulse waveforms:

$$\begin{aligned}\tilde{I}_{\mathrm{PD}}[t] &= \mathrm{cat}_{\rightarrow}\,\mathrm{cat}_{\downarrow}\begin{pmatrix}\tilde{I}_o(t_i^{(1)}, t_p^{(1)}) & \tilde{I}_o(t_i^{(1)}, t_p^{(2)}) & \cdots & \tilde{I}_o(t_i^{(1)}, t_p^{(m)}) \\ \tilde{I}_o(t_i^{(2)}, t_p^{(1)}) & \tilde{I}_o(t_i^{(2)}, t_p^{(2)}) & \cdots & \tilde{I}_o(t_i^{(2)}, t_p^{(m)}) \\ \vdots & \vdots & \ddots & \vdots \\ \tilde{I}_o(t_i^{(n)}, t_p^{(1)}) & \tilde{I}_o(t_i^{(n)}, t_p^{(2)}) & \cdots & \tilde{I}_o(t_i^{(n)}, t_p^{(m)})\end{pmatrix} \\ &\rightleftharpoons \mathrm{cat}_{\rightarrow}\,\mathrm{cat}_{\downarrow}\begin{pmatrix}\tilde{I}_o(\lambda_1, t_p^{(1)}) & \tilde{I}_o(\lambda_1, t_p^{(2)}) & \cdots & \tilde{I}_o(\lambda_1, t_p^{(m)}) \\ \tilde{I}_o(\lambda_2, t_p^{(1)}) & \tilde{I}_o(\lambda_2, t_p^{(2)}) & \cdots & \tilde{I}_o(\lambda_2, t_p^{(m)}) \\ \vdots & \vdots & \ddots & \vdots \\ \tilde{I}_o(\lambda_n, t_p^{(1)}) & \tilde{I}_o(\lambda_n, t_p^{(2)}) & \cdots & \tilde{I}_o(\lambda_n, t_p^{(m)})\end{pmatrix}\end{aligned} \tag{6.17}$$

where $\mathrm{cat}_{\rightarrow}$ and $\mathrm{cat}_{\downarrow}$ mean horizontal and vertical concatenations, respectively. Each $\tilde{I}_o(t_i^{(n)}, t_p^{(m)})$ expresses the nth spectral (spatial) pixel at the mth pulse (line image). Applying Eq. (6.16) to Eq. (6.13),

$$\tilde{I}_o(t_i^{(n)}, t_p^{(m)}) \propto I_b(t_i^{(n)}, t_p^{(m)}) + I_i(t_i^{(n)}, t_p^{(m)}) \cdot \cos\left[\frac{4\pi\Delta L \cdot DL_f}{t_i^{(n)} + D\lambda_c L_f} + \Delta\varphi_c(t_i^{(n)}, t_p^{(m)})\right] \tag{6.18}$$

Therefore, the time stretched temporal waveform corresponding to each line scan image consists of two features [19]: One is $I_b(t_i^{(n)}, t_p^{(m)})$, a temporal envelope of the time stretched optical pulse at baseband frequencies. The amplitude of this envelope corresponds to the temporal shape of the optical pulse and its deviations caused by the object transmission as in brightfield microscopy. It provides information about optical loss, i.e., light absorption and scattering caused by surface roughness, granularity, and inner cell organelle complexity.

The second term in Eq. (6.18) (with cosine component) is a fast oscillating fringe, caused by the spectral interference of the pulses multiplexed between the sample and the reference arms in the Michelson interferometer. This term can be separated by a bandpass filter, and its envelope can be derived by a nonlinear envelope detection technique. Here we used a moving minimum/maximum filter to extract the envelope. After normalization to the envelope, the cosine component

$$I_c(t_i^{(n)}, t_p^{(m)}) = \cos\left[\frac{4\pi \Delta L \cdot DL_f}{t_i^{(n)} + D\lambda_c L_f} + \Delta\varphi_c(t_i^{(n)}, t_p^{(m)})\right] \tag{6.19}$$

is used for calculation of the object phase shift, $\Delta\varphi_c(x, t_p)$. The first term in cosine causes the interferogram fringe pattern. Since $t_i \ll D\lambda_c L_f$, it can be approximated as

$$\frac{4\pi DL_f \Delta L}{t_i + D\lambda_c L_f} \approxeq -f_i t_i + \varphi_{i0} \tag{6.20}$$

where φ_{i0} is an initial phase constant, f_i is the fringe frequency:

$$f_i \approxeq \frac{4\pi \Delta L}{\lambda_c^2 DL_f} \tag{6.21}$$

As seen in Fig. 6.2b, the fringe frequency, f_i, in our setup is about 4.7 GHz determined by the optical path length mismatch between the interferometer arms.

The instantaneous phase of $I_c(t_i^{(n)}, t_p^{(m)})$ can be readily retrieved from its analytic representation given by Hilbert transform, $\mathcal{H}$:

$$\begin{aligned} \angle I_c(t_i^{(n)}, t_p^{(m)}) &= \arg\left[I_c(t_i^{(n)}, t_p^{(m)}) + j \cdot \mathcal{H}\{I_c(t_i^{(n)}, t_p^{(m)})\}\right] \\ &= \frac{4\pi DL_f \Delta L}{t_i^{(n)} + D\lambda_c L_f} + \Delta\varphi_c(t_i^{(n)}, t_p^{(m)}) \end{aligned} \tag{6.22}$$

Here arg means the argument of a complex number. A one-dimensional phase unwrapping algorithm followed by background phase removal gives the object phase shift,

$$\begin{aligned} \Delta\varphi_c(t_i^{(n)}, t_p^{(m)}) = \text{unwrap}\{&\angle I_c(t_i^{(n)}, t_p^{(m)}) \\ &- \angle I_c(t_i^{(n)}, t_p^{(\text{empty})})\} \end{aligned} \tag{6.23}$$

where $t_p^{(\text{empty})}$ corresponds to an empty pulse when no cell is in the field of view, i.e., background phase. The unwrapping algorithm used in our processing acts when the absolute phase difference between two consecutive samples of the signal is greater than or equal to π radians, and adds multiples of 2π to the following samples in order to bring the consecutive samples phase difference in the acceptable range of $-\pi$ to π.

To perform combined quantitative phase and loss imaging, the phase derived by Hilbert transformation should be corrected to eliminate the artifacts caused by the intensity variations induced by the passing cells. Most cells of interest in clinical or industrial applications have a diameter 3–40 μm, when suspended in fluid. Given the field of view and the period of the interrogation rainbow pulses are 130 μm and 27 ns, respectively, the time duration of the instantaneous intensity change induced by the single cells in each laser pulse is about 0.6–8.3 ns, which will bring in frequency components up to about 1.6 GHz. Compared to the higher frequency components at 4.7 GHz corresponding to the interference fringes, the frequency of intensity variations is small (<1.6 GHz), and in this scenario, our method remains robust to separate the two electrical spectral components for optical loss and phase.

6.2.6 Cell Transmittance Extraction

One of the greatest advantage of TS-QPI is its ability to extract the cell transmittance, $T_s(\lambda)$, without prior knowledge of the transmittance of the solution, $T_s(\lambda)$, that of the beam-splitter, $T_b(\lambda)$, and the reflectance of substrate of the microfluidic channel, $R_m(\lambda)$. During measurements when there is no cell in the field of view (empty frames), Eq. (6.15) becomes

$$I_i(\lambda, t_p^{(\text{empty})}) = -2[1 - T_b(\lambda)]T_b(\lambda)R_m(\lambda) \cdot T_s(\lambda)|\tilde{E}_{in}(\lambda, t_p^{(\text{empty})})|^2 \tag{6.24}$$

In addition, the signal from only the reference arm can be recorded by blocking the sample arm:

$$I_r(\lambda, t_p) = [1 - T_b(\lambda)]^2 R_m(\lambda)|\tilde{E}_{in}(\lambda, t_p)|^2 \tag{6.25}$$

Combining Eqs. (6.14), (6.24), and (6.25), and assuming that the input electric field pulse shape, $|\tilde{E}_{in}(\lambda, t_p)|$, is invariant to t_p, the cell transmittance can be derived as

$$T_c(\lambda) = \frac{-2\sqrt{I_r(\lambda, t_p) \cdot (I_b(\lambda, t_p) - I_r(\lambda, t_p))}}{I_i(\lambda, t_p^{(\text{empty})})} \tag{6.26}$$

Please note that the values of $I_r(\lambda, t_p)$, $I_b(\lambda, t_p)$, and $I_i(\lambda, t_p^{(\text{empty})})$ are directly measured by TS-QPI, and no prior knowledge of $T_b(\lambda)$, $T_s(\lambda)$, $R_m(\lambda)$, and $|\tilde{E}_{in}(\lambda, t_p)|$ is needed to calculate the cell transmittance.

6.2.7 Image Reconstruction

We reconstruct both quantitative brightfield and phase-contrast images simultaneously from single-shot frequency-multiplexed interferometric measurements. The envelope and phase of the time-domain signal $\tilde{I}_o(t_i^{(n)}, t_p^{(m)})$ were firstly mapped into series of spatial information $\tilde{I}_o(x^{(n)}, t_p^{(m)})$, forming a line scanning brightfield image and phase contrast image, illuminated by the optical pulse at time t_p. This is because within each optical pulse, the spatial information is mapped one-to-one into spectral domain, $x^{(n)} \rightarrow \lambda_n$, and spectrum is stretch in time, $\lambda_n \rightarrow t_i^{(n)}$, where $t_i^{(n)}$ is the relative group delay time of each frequency component within a pulse with respect to the central wavelength. These line-scan images based on $\tilde{I}_o(x, t_p^{(1)})$, $\tilde{I}_o(x, t_p^{(2)})$, $\tilde{I}_o(x, t_p^{(3)})$, ... were then cascaded into a two-dimensional image corresponding to $\tilde{I}_o(x, y)$, where the second dimension y is the spatial mapping of time lapse based on object flow speed.

The optical path length difference image can be calculated by the phase shift line scans as

$$\mathrm{OPD}(x^{(n)}, y^{(m)}) = \frac{\lambda(t_i^{(n)}, t_p^{(m)})}{2\pi} \Delta\varphi_c(t_i^{(n)}, t_p^{(m)}) \tag{6.27}$$

On the other hand, if the axial thickness of the cell at reconstructed image pixel (x, y) is $d(x, y)$,

$$\mathrm{OPD}(x, y) = 2[n_{\mathrm{cell}}(x, y) - n_{\mathrm{solution}}(x, y)] \cdot d(x, y) \tag{6.28}$$

in which n_{cell} and n_{solution} are the refractive indices of the cell and the surrounding buffer solution, respectively. The factor 2 is to account for the fact that each wavelength component passes the cell twice in Michelson interferometer.

If we integrate Eq. (6.28) over the area of the cell, we can derive an average refractive index contrast for the cell, which corresponds to the average protein concentration of the cell:

$$\overline{\Delta n_{\mathrm{cell}}} = \overline{n_{\mathrm{cell}} - n_{\mathrm{solution}}} = \frac{\iint_{\mathrm{cell}} \mathrm{OPD}(x, y)dxdy}{2\iint_{\mathrm{cell}} t(x, y)dxdy} \tag{6.29}$$

where $\iint_{\mathrm{cell}} t(x, y)dxdy$ is the volume of the cell obtained from its lateral diameter, d, as $V \approx \pi d^3/6$.

The unit net change of intensity envelope variations induced by the cell is obtained from the amplitude of the slowly varying envelope feature of the interferogram as

$$\Delta I_b(\lambda, t_p) = \frac{T_b^2(\lambda)T_s^2(\lambda)(1 - T_c^2(\lambda, t_p))}{[1 - T_b(\lambda)]^2 + T_b^2(\lambda)T_s^2(\lambda)} \tag{6.30}$$

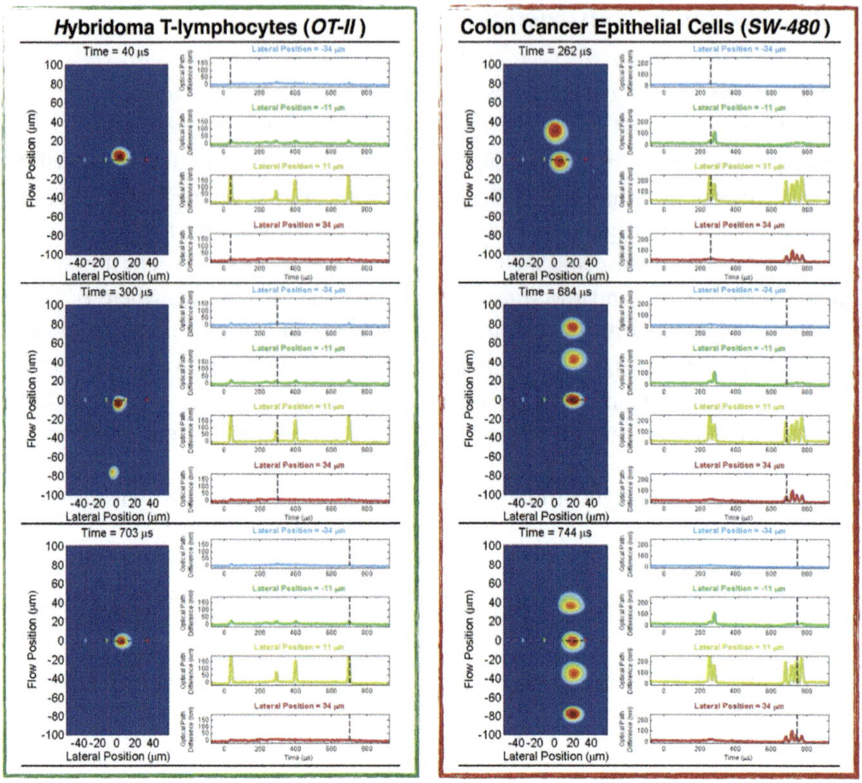

Fig. 6.4 *Left*: screenshots of the video of *OT-II* hybridoma T-lymphocytes flowing in a microfluidic channel; The cells are aligned at the center of the channel by hydrodynamic focusing. Optical path difference measured at four of the interrogation points on the rainbow flash is shown as a function of time in the *right panels*. *Right*: screenshots of the video of *SW-480* colon cancer epithelial cells flowing in a microfluidic channel; The cells are aligned at the center of the channel by hydrodynamic focusing. Optical path difference measured at four of the interrogation points on the rainbow flash is shown as a function of time in the *right panels*

It gives the temporal and spatial information of the combined effects from absorption and scattering:

$$I_{\text{loss}}(x^{(n)}, y^{(m)}) = \Delta I_b(\lambda_n, t_p^{(m)}) \tag{6.31}$$

It gives the temporal and spatial information of the combined effects from absorption and scattering (Fig. 6.4).

6.3 Image Processing Pipeline

6.3.1 *Feature Extraction*

The decomposed components of sequential line scans form pairs of spatial maps, namely, optical phase and loss images (Fig. 6.5). These images are used to obtain biophysical fingerprints of the cells [29, 118]. With domain expertise, raw images are fused and transformed into a suitable set of biophysical features, listed in Table 6.2, which the deep learning model further converts into learned features for improved classification.

The feature extraction operates on optical phase and loss images simultaneously, including object detection, segmentation, and feature measurement, as well as clump identification, noise suppression, etc. As an example of the expert designed features, the average refractive index, used as a measure of protein concentration [119], is obtained by dividing the integral of the optical path length by the cell volume. Since cells in suspension relax to a spherical shape (due to surface tension) [93, 94], an independent measure of cell diameter can be obtained from its lateral dimension for volume estimation.

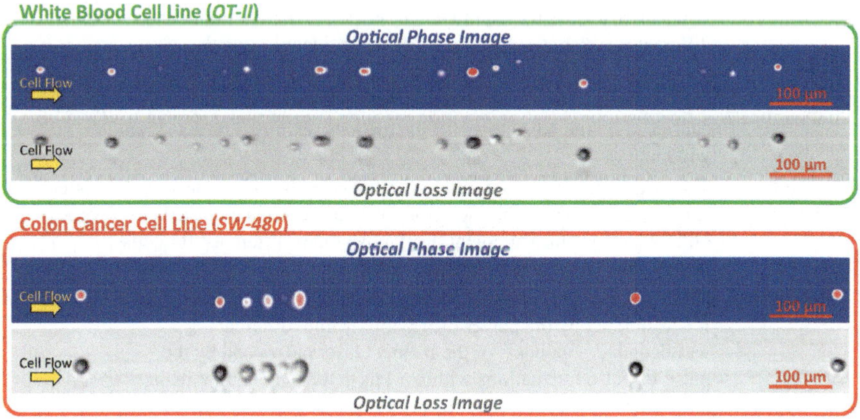

Fig. 6.5 Quantitative optical phase and loss images of *OT-II* (*blue*) and *SW-480* (*green box*) cells; The optical loss images of the cells are affected by the attenuation of multiplexed wavelength components passing through the cells. The attenuation itself is governed by the absorption of the light in cells as well as the scattering from the surface of the cells and from the internal cell organelles. The optical loss image is derived from the low frequency component of the pulse interferograms. The optical phase image is extracted from the analytic form of the high frequency component of the pulse interferograms using Hilbert Transformation, followed by a phase unwrapping algorithm. Also, supplementary Videos 1 and 2 show measurements of cell-induced optical path length difference by TS-QPI at four different points along the rainbow for *OT-II* and *SW-480*, respectively

Table 6.2 List of extracted features

Feature name	Description	Category
Diameter-RB	Diameter along the interrogation rainbow. It is insensitive to flow rate fluctuation. For higher accuracy, it is calibrated by the spatial nonuniform distribution of rainbow wavelengths	Morphology
Diameter-FL	Diameter along the flow direction. It is sensitive to flow rate fluctuation, but can be a candidate parameter for monitoring flow speed and channel condition	Morphology
Tight area	Total number of pixels in the segmented region in the phase image	Morphology
Perimeter	Total number of pixels around the boundary of each segmented region	Morphology
Circularity	$4\pi\,\text{Area}/\text{Perimeter}^2$	Morphology
Major axis	Considering the cell as elliptical in lateral imaging plane, the length of the major axis of the ellipse with a normalized second central moment same as the cell	Morphology
Orientation	Angle between the flow direction and the major axis of the cell elliptical shape	Morphology
Loose area	Total number of pixels in the expanded segmented region for measurement of the pixel intensities	Morphology
Median radius	The median distance of any pixel in the object to the closest pixel outside of the object	Morphology
OPD-1	Integrated optical path length difference within the entire segmented area (cell), calibrated by the power distribution within different wavelength components of the incident laser pulses	Optical phase
OPD-2	Integrated optical path length difference within the entire segmented area (cell). In addition to the calibration of OPD-1, it is calibrated by the pulse-to-pulse fluctuations within a 1 μs detection window	Optical phase
Refractive index	The mean refractive index difference between the object and the surrounding liquid (buffer solution), which is calculated based on OPD-2 and size measurement. Refractive index difference for cells is proportional to their protein concentration	Optical phase
Absorption-1	Mean absorption coefficient within the entire segmented area (cell). It is calibrated by the power distribution within different wavelength components of the incident laser pulses and by the pulse-to-pulse fluctuations within a 1 μs detection window. This parameter corresponds to an absorption-dominant model for the cell	Optical loss
Absorption-2	Mean absolute absorption coefficient within the entire segmented area (cell). It is calibrated by the power distribution within different wavelength components of the incident laser pulses and by the pulse-to-pulse fluctuations within a 1 μs detection window. This parameter corresponds to an absorption-dominant model for the cell	Optical loss
Scattering-1	Mean optical loss within the entire segmented area (cell). It is calibrated by the power distribution within different wavelength components of the incident laser pulses and by the pulse-to-pulse fluctuations within a 1 μs detection window. This parameter corresponds to a scattering-dominant model for the cell	Optical loss
Scattering-2	Mean absolute optical loss within the entire segmented area (cell). It is calibrated by the power distribution within different wavelength components of the incident laser pulses and by the pulse-to-pulse fluctuations within a 1 μs detection window. This parameter corresponds to a scattering-dominant model for the cell	Optical loss

The high-content image analysis and cell screening pipeline is implemented by combining multiple informatics tools, namely CellProfiler for image processing [33, 120], MySQL/MangoDB for database, Matlab for machine learning, and Javascript for interactive visualization. First of all, image noise reduction and smoothing have been performed, which can remove artifacts that are smaller than optical resolution limit. For object segmentation, we use the Otsu's thresholding method. Once objects are identified in the image, morphology of each single cell can be described by area, diameter, uniformity, aspect ratio, perimeter, number of surrounding clumped cells, etc.

The capability to identify clumped cells from single large cells greatly reduces the misclassification rate in imaging flow cytometry compared to traditional flow cytometry. Intensity peaks of pixel brightness within each object are used to distinguish clumped objects. The object centers are defined as local intensity maxima in the smoothed image. Retaining outlines of the identified objects helps validate and visualize the algorithm. In the next step, we discard the objects touching the borders of the image, i.e., the edges of the field of view and data acquisition time window. However, the chance of cells showing up at the edges is very low due to hydrodynamic focusing. We are also capable of excluding dust, noise, and debris by neglecting the objects that are too small or their aspect ratio is too extreme to be a cell.

6.3.2 Multivariate Features Enabled by Sensor Fusion

In feature extraction, one of the most important advantages of optical loss and phase fusion is its robustness and insensitivity to axial defocusing [121] caused by the limited depth-of-focus of the objective lens and variations of the cell alignment in microfluidic channel. Diffracted photons have little chance to be influential in phase images. This makes the size measurements in optical phase images relatively accurate and consistent, more suitable than direct size measurements in optical loss images for extraction of scattering and absorption features. Among different features, size measurement is particularly important as it is used by itself in many technologies [87, 122–124].

The large data set captured by TS-QPI provides sufficient statistical characteristics for cell analysis based on biophysical features. Since cells from even the same line or tissue exhibit variations in size, structure, and protein expression levels [32, 125, 126], high accuracy classification can only be achieved by a model tolerant to these intrinsic variations On the other hand, the feature extractor must reflect the intricate and tangled characteristics caused by extrinsic variations, e.g., drug treatment [112], cell cycles, rare cell types, labeling, and transcription rate [127].

A total of 16 features are chosen among the features extracted from fusion of optical phase and loss images of each cell. Features that are highly correlated do not provide unique information. Pairwise correlation matrix among these features is shown as a heat map in Fig. 6.6a. Diagonal elements of the matrix are correlation of

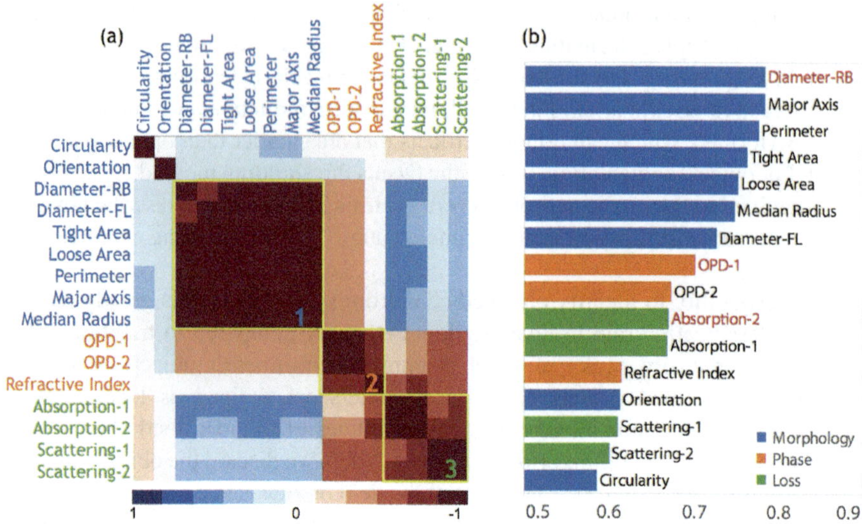

Fig. 6.6 (**a**) Pairwise correlation matrix visualized as a heat map. The map depicts the correlation between all major 16 features extracted from the quantitative images. Diagonal elements of the matrix represent correlation of each parameter with itself, i.e., the autocorrelation. The subsets in *box 1*, *box 2*, and *box 3* show high correlation because they are mainly related to morphological, optical phase, and optical loss feature categories, respectively. (**b**) Ranking of biophysical features based on their AUCs in single-feature classification. *Blue bars* show performance of the morphological parameters, which includes diameter along the interrogation rainbow, diameter along the flow direction, tight cell area, loose cell area, perimeter, circularity, major axis length, orientation, and median radius. As expected, morphology contains most information, but other biophysical features can contribute to improved performance of label-free cell classification. *Orange bars* show optical phase shift features, i.e., optical path length differences and refractive index difference. *Green bars* show optical loss features representing scattering and absorption by the cell. The best performed features in these three categories are marked in *red*

each feature with itself, i.e., the autocorrelation. The subset of the features in Box 1 shows high correlation among morphological features. Also, the subset features in Box 2 and 3 are correlated as they are mainly related to optical phase shift and optical loss, respectively.

As a representation of our biophysical features in classification, Fig. 6.6b shows classification accuracy based on each single feature arranged in descending order. The features are color coded into three categories: morphology, optical phase, and optical loss, to describe the main type of information provided by each. The figure provides valuable insight into the relative importance of each category of cell features and suggests that morphological features carries the most information about cells, but at the same time, significant additional information is contained in optical phase and loss measurements.

6.3.3 *System Calibration*

To calibrate the imaging system and image processing pipelines for size measurement, 5 μm polystyrene beads (from Polysciences, Inc.) with NIST traceable particle size standards were analyzed. Size measurement of the polystyrene beads had a distribution with 5.06 μm expected mean and 0.5 μm standard deviation. The broadened standard deviation was within the range of optical resolution limit and was caused mainly by performing object recognition on resolution limited images. Due to limited optical resolution of the setup, the edges of bead or cell are blurred, generating distribution of point spread functions in optical phase and loss images outside of the cell boundaries. In order to maximize the accuracy in morphological, phase, and loss measurements, after object segmentation we expanded the object boundaries by 2.5 μm (optical resolution of the setup measured by knife-edge method), which serve as loose boundaries, indicating the area within which the pixel intensities are measured and integrated in phase and loss images.

6.4 Conclusion

The multiplexed biophysical features thus lead to information-rich hyper-dimensional representation of the cells for label-free classification with high statistical precision. In the next chapter, we further improved the accuracy, repeatability, and the balance between sensitivity and specificity of our label-free cell classification by a novel machine learning pipeline, which harnesses the advantages of multivariate supervised learning, as well as unique training by evolutionary global optimization of receiver operating characteristics (ROC).

Part IV
Big Data and Artificial Intelligence

Chapter 7
Big Data Acquisition and Processing in Real-Time

Coherent-STEAM is a quantitative phase microscopy technique for label-free analysis of up to 100,000 cells per second in flow. Here, we introduce a data acquisition scheme that enables interruptionless storage of Coherent-STEAM cell images. Our proof of principle demonstration is capable of saving 10.8 TB of cell images in an hour, i.e., pictures of every single cell in 2.7 mL of a sample.

7.1 Introduction

In Coherent-STEAM system, the output signal of the photodetector usually has a very large bandwidth in the order of a few GHz. Based on Nyquist theorem, an analog-to-digital converter (ADC) with a sampling rate of at least twice the bandwidth is required to capture this signal without aliasing. If we want to capture images of every single cell in a sample, all of the ADC output samples should be recorded on a storage unit. For example, to capture and process the data for the Coherent-STEAM setup, the minimum sampling rate of the photodetector signal is 12 GS/s. If the bit-depth of the ADC is 8 bits, the acquisition system should handle storage of 12 GB of data per second. To ease the storage requirement by a few times, we purpose analog preprocessing of Coherent-STEAM data. Our technique is based on using telecommunication radio-frequency (RF) components to convert the photodetector output signal into a set of lower bandwidth signals with only the quantitative phase and intensity information, so that the sampling rate of the required ADCs can be much smaller. As a result, the storage units can handle logging these lower rate signals at real-time. Also, by parallel processing of the stored data, we would be able to retrieve the phase and intensity images in real-time and use them for cell sorting.

A. Mahjoubfar et al., *Artificial Intelligence in Label-free Microscopy*,
DOI 10.1007/978-3-319-51448-2_7

7.2 Technical Description of the Acquisition System

We have previously [19] shown that if the arms' length mismatch in Coherent-STEAM interferometer is chosen long enough, one can see two separate features in the spectrum of the system output corresponding to intensity and phase components of the cells (Fig. 7.1). By filtering out the high-frequency features and down-converting them to baseband, we should be able to reconstruct intensity and phase images of the cells.

We purpose a quadrature phase demodulation scheme to perform the analog preprocessing on Coherent-STEAM signals (Fig. 7.2). First, the photodetector output signal is bandpass filtered and split into two paths. These signals are mixed with two sinusoidal signals that are 90° phase-shifted with respect to each other. The frequency of the sinusoidal signals is approximately at the center of the high frequency features of Coherent-STEAM setup, which is set by the arms' length mismatch (in our example about 5 GHz). Mixers shift the high-frequency component containing the phase and intensity information to lower frequencies close to baseband forming in-phase, $I(t)$, and quadrature, $Q(t)$, demodulation components (I/Q demodulation).

Finally, the baseband components, which now contain the sample's phase and intensity information, can be filtered out and digitized with two ADCs (Fig. 7.3) that have a considerably smaller sampling rate than what was required before the down-conversion. In our demonstration, the sampling rate was reduced from 12 GS/s to 1.5 GS/s. In addition, since the outputs are mixed with 90° phase-shifted sinusoidal signals, the phase and intensity of the signal can be respectively derived by simple calculations as

$$\Delta\varphi = \text{unwrap}\left(\arg\left(\frac{I(t)}{Q(t)}\right)\right), \tag{7.1}$$

$$\Delta I = \sqrt{I(t)^2 + Q(t)^2}, \tag{7.2}$$

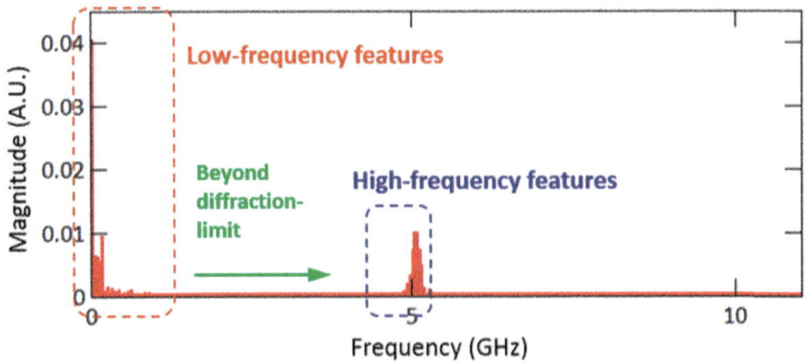

Fig. 7.1 Spectral components of Coherent-STEAM signal. For a Coherent-STEAM setup with long enough arms' length mismatch the spectrum of the output signal shows two separate spectral bands. The low frequency components correspond to the intensity of the sample, while the high frequency components contain the phase information in addition to the intensity information of the cells

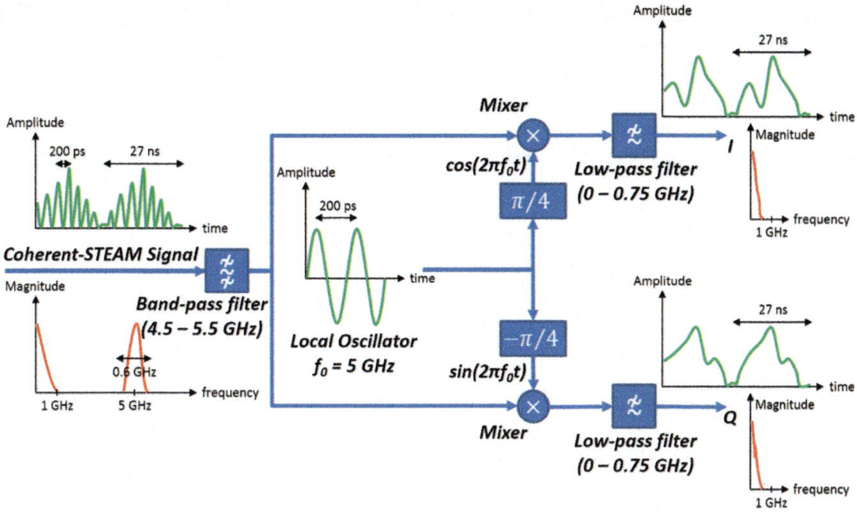

Fig. 7.2 Analog preprocessing of Coherent-STEAM signal. The analog signal processing system for reducing the data rate of Coherent-STEAM is essentially a quadrature down-conversion unit. I and Q outputs and their corresponding spectra show that the down-conversion is effective in reducing the bandwidth and the required sampling rate

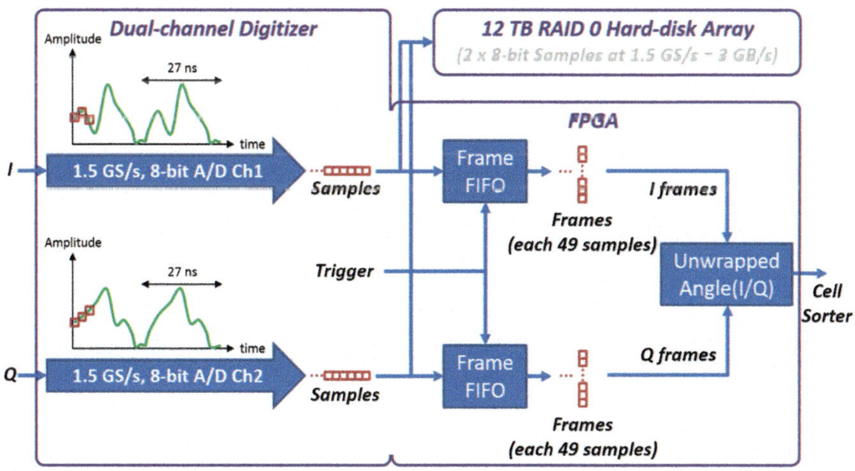

Fig. 7.3 Digital signal processing system for acquisition of analog preprocessing unit outputs. This system is built with simple blocks such as argument calculator, unwrapper, and first in, first outs (FIFOs), which can be performed in real-time

where $I(t)$ and $Q(t)$ are the in-phase and quadrature-phase outputs of the analog preprocessing unit as shown in Fig. 7.2.

7.3 Big Data Acquisition Results

We tested the applicability of our method with a preliminary setup. To generate 90° phase-shifted sinusoidal signals, we used a signal generator connected to a 90° hybrid coupler. The $I(t)$ and $Q(t)$ outputs of the analog signal processing system are captured with two 1.5 GS/s analog-to-digital converters. These signals are down-converted in frequency domain and about eight times slowed-down in time compared to the original Coherent-STEAM output (Fig. 7.2). This down-conversion happens for consecutive line images at real-time. Also, with careful design, both channels can have the same group delay, and edges of the pulses in two channels can align in time.

Figure 7.4 shows a few intensity and phase cell images captured by our continuously recording big data acquisition system. The setup can acquire 10.8 TB of these images over a course of an hour-long experiment, which corresponds to pictures of every single cell in 2.7 mL of the suspended cells sample. The storage unit is a RAID 0 array of hard disk drives that are written and read in parallel to provide superior access speed. To capture larger data sizes, the system can be easily expanded by increasing the number of hard disk drives in the array or using larger hard drives.

Also, the required digital signal processing for derivation of sample phase-shift from the outputs of the analog signal processing system, $I(t)$ and $Q(t)$, can be

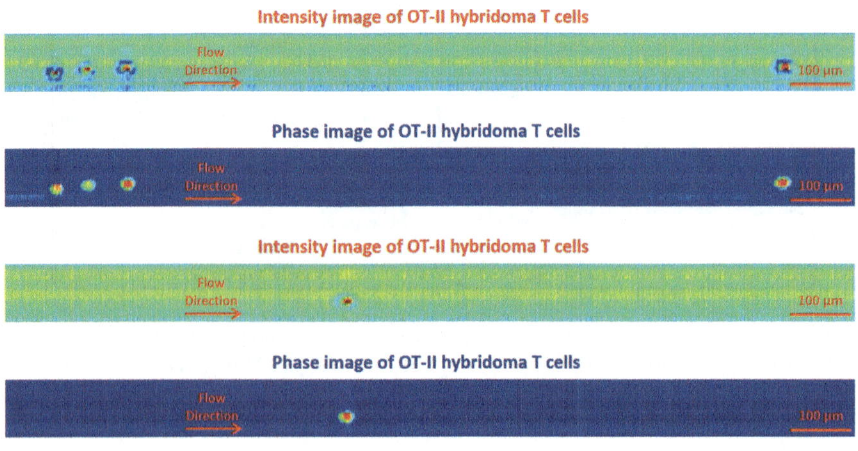

Fig. 7.4 Sample images acquired by the analog preprocessing system. Both phase and intensity images for two different sets of OT-II hybridoma T cells in flow are shown

easily implemented on an FPGA. Figure 7.3 shows a suggested design for such an FPGA unit. One can see that it only requires implementation of basic blocks such as argument calculator, unwrapper, and first in, first outs (FIFOs). This is a direct result of transferring the cumbersome and calculation intensive operations of the phase recovery algorithm (such as Hilbert transformation) to the analog preprocessing unit. This way, the FPGA output signal can be directly used to control a cell sorter in our label-free imaging flow cytometer.

7.4 Conclusion

In summary, we used quadrature phase demodulation technique to reduce the sampling rate required for capturing Coherent-STEAM signals, decrease the amount of data that is generated, and facilitate the data processing. As a proof of principle, we showed an acquisition system capable of continuously recording 10.8 TB of phase and intensity images.

Chapter 8
Deep Learning and Classification

As demonstrated in previous chapters, our TS-QPI system captures quantitative optical phase and intensity images and extracts multiple biophysical features of individual cells. In this chapter, we use these biophysical measurements to form a hyperdimensional feature space in which supervised learning is performed for cell classification. We show that TS-QPI not only overcomes the throughput issue in cellular imaging, but also improves label-free diagnosis by integration of sensing multiple biophysical features. We also compare various learning algorithms including artificial neural network, support vector machine, logistic regression, and a novel deep learning pipeline, which adopts global optimization of receiver operating characteristics. As a validation of the enhanced sensitivity and specificity of our system, we show classification of white blood T-cells against colon cancer cells, as well as lipid accumulating algal strains for biofuel production. This system opens up a new path to data-driven phenotypic diagnosis and better understanding of the heterogeneous gene expressions in cells.

8.1 Background

Deep learning refers to methods for extraction of patterns and knowledge from rich multidimensional datasets. While it is extensively used for image recognition and speech processing, its application to label-free classification of cells has not been exploited.

Here, we further improved the accuracy, repeatability, and the balance between sensitivity and specificity of label-free phenotypic screening by a novel machine learning pipeline, which harnesses the advantages of multivariate supervised learning, as well as unique training by evolutionary global optimization of receiver operating characteristics (ROC). To demonstrate sensitivity, specificity, and accuracy of multi-feature label-free flow cytometry using our technique, we classified

A. Mahjoubfar et al., *Artificial Intelligence in Label-free Microscopy*,
DOI 10.1007/978-3-319-51448-2_8

(1) *OT-II* hybridoma T-lymphocytes and *SW-480* colon cancer epithelial cells, and (2) *Chlamydomonas reinhardtii* algal cells (herein referred to as Chlamydomonas) based on their lipid content, which is related to the yield in biofuel production. Our preliminary results show that compared to classification by individual biophysical parameters, our label-free hyperdimensional technique improves the detection accuracy from 77.8% to 95.5%, or in other words, reduces the classification inaccuracy by about five times.

8.2 Machine Learning

Neural networks are a flexible and powerful bioinspired learning model, which perform layers of nonlinear feature transformations, learned from the training data [128–130]. The transformations morph the input data with weighted sums and nonlinear activation functions into feature spaces more suitable for classification. Shown in Fig. 8.1 is a unique feedforward neural network learning model that is globally trained by the objective of improving receiver operating characteristic (ROC). The learning algorithm introduced here maximizes the area under ROC curve (AUC), which is a global indicator of the classifier performance on the entire

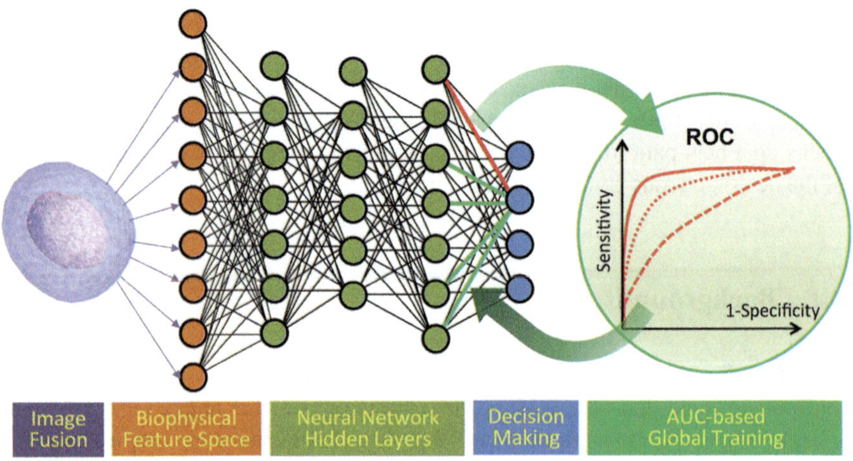

Fig. 8.1 Machine learning pipeline. Information of quantitative optical phase and loss images are fused to extract multivariate biophysical features of each cell, which are fed into a fully connected neural network. The neural network maps input features by a chain of weighted sum and nonlinear activation functions into learned feature space, convenient for classification. This deep neural network is globally trained via area under the curve (AUC) of the receiver operating characteristics (ROC). Each ROC curve corresponds to a set of weights for connections to an output node, generated by scanning the weight of the bias node. The training process maximizes AUC, pushing the ROC curve toward the upper left corner, which means improved sensitivity and specificity in classification

training dataset [131–133]. The global training of the neural network, although computationally costly, results in a classifier more robust, repeatable, and insensitive to imbalance among classes. For the purpose of end-to-end supervised learning with AUC whose gradient is not well behaved, we employed the heuristic genetic algorithm (GA), which is resilient to discontinuities of the cost function and being trapped in local minima during optimization.

The network is composed of multiple hidden layers, which automatically learn representations of the data at different levels of abstraction, and thus is considered a form of deep learning [134, 135]. Each layer performs a linear combination on its inputs from the previous layer and operates a nonlinear function on the weighted sums. The output of the node j in layer $l+1$, denoted by $z_j^{(l+1)}$, is generated from inputs $x_1, x_2, \ldots, x_N$ as

$$z_j^{(l+1)} = \mathrm{h}\left(a_j^{(l+1)}\right) = \mathrm{h}\left(\sum_{i=0}^{N_l} \omega_{ji}^{(l)} x_i^{(l)}\right) \tag{8.1}$$

$$l = 0, \ldots, L; \quad j = 1, \ldots, N_{l+1}$$

Here $a_j^{(l+1)}$ is the linear combination of inputs, and $\omega_{ji}^{(l)}$ are the weights of the linear combination. The summation runs over N_l, the total number of nodes in the layer l, and L is the total number of hidden layers. $x_0^{(l)}$ is the bias node in layer l, conventionally 1. Some popular choices for the nonlinear activation function $\mathrm{h}(\cdot)$ include logistic sigmoid function $\mathrm{h}(a) = 1/(1+\exp(-a))$, hyperbolic tangent function $\tanh(a)$, and commonly used in deep learning, rectified linear unit (ReLU) $\mathrm{h}(a) = \max(0, a)$. In our learning model, we use ReLU, which typically speeds up the supervised learning process of deep neural network by inducing sparsity and preventing gradient vanishing problem.

For a trained classifier in hyperspace, receiver operating characteristics (ROC) curve describes the sensitivity and specificity of a classifier collection that includes nonlinear classifiers scaled in the direction of their normal vector field. In a deep learning network, this is equivalent to shifting the weight of the bias node in the last hidden layer. ROC highlights the trade-off between sensitivity and specificity (Fig. 8.1), and the area under ROC (AUC) provides a quantitative robust measure of classifier performance [136–139]. Choosing a large value for the weight of the bias node results in high sensitivity; however this sacrifices the specificity leading to large number of false positives. As a way to visualize the impact of the threshold on classification accuracy, a classifier that accurately separates the classes will have an ROC curve that approaches the upper left corner. Conversely, a random guess, corresponding to accuracy of 50% in binary classification will have an ROC that is a diagonal line. The AUC parameter serves as an effective analysis metric for finding the best classifier collection and has been proven to be advantageous over the mean square error for evaluating learning algorithms [140].

To prevent overfitting in our deep learning model, we added a regularization term to the AUC-based cost function. Our regularization term is defined as mean square of all the network weights, excluding the weight of the bias nodes. Therefore, the overall cost function, cost(ω), that is minimized by the genetic algorithm is

$$\begin{aligned}\mathrm{cost}(\omega) = {} & (1 - \mathrm{AUC}(\omega)) \\ & + \lambda \frac{\sum_{l=0}^{L} \sum_{j=1}^{N_l+1} \sum_{i=1}^{N_l} (\omega_{ji}^{(l)})^2}{\sum_{l=0}^{L} \sum_{j=1}^{N_l+1} \sum_{i=1}^{N_l} 1}\end{aligned} \tag{8.2}$$

where λ is the regularization parameter, which controls the trade-off between overfitting (variance) and underfitting (bias).

8.3 Applications

8.3.1 *Blood Screening: Demonstration in Classification of* OT-II *and* SW-480 *Cells*

In previous chapter, the multiplexed biophysical features extracted from ultra-fast cellular imaging using TS-QPI lead to information-rich hyper-dimensional representation of the cells. In contrast to single-feature approaches [31, 87, 123], our label-free cell classification enabled by TS-QPI and multivariate analysis offers considerable improvements in detection sensitivity and accuracy for cancer diagnosis (Fig. 8.2). To demonstrate the application in circulating tumor cell (CTC) detection, we used *OT-II* hybridoma T cells as a model for normal white blood cells and *SW-480* epithelial colon cancer cells. The features described in Table 6.2 were measured by our TS-QPI system for the aforementioned cells. Figure 8.3 shows three of these features in a three-dimensional (3D) scatter plot, attributed to size, protein concentration, and attenuation. The 2D projections on the three orthogonal planes are also shown. It is clear that additional dimensions improve distinguishment among different cell types compared to individual features.

A five-fold cross-validation methodology is applied on the dataset to split data points into training, validation, and test subsets. Figure 8.2a shows progress in label-free classification depicted by balance accuracy as the learning model evolves over GA generations. Blue curve shows the classification balanced accuracy of the test dataset using all sixteen biophysical features extracted from the TS-QPI images. To highlight the improvement by hyperdimensional feature space of TS-QPI, we also show the balanced accuracy curves based on several single features: cell diameter for morphology, integral of cell's optical path difference for optical phase information, and cellular absorption for optical loss in near-infrared window. Although these three biophysical features individually perform the highest accuracy among morphology, optical phase, and optical loss groups respectively, as previously shown in Fig. 6.6b, our multivariate deep learning classifier outperforms them. In addition, receiver operating characteristic (ROC) curves for each fold are

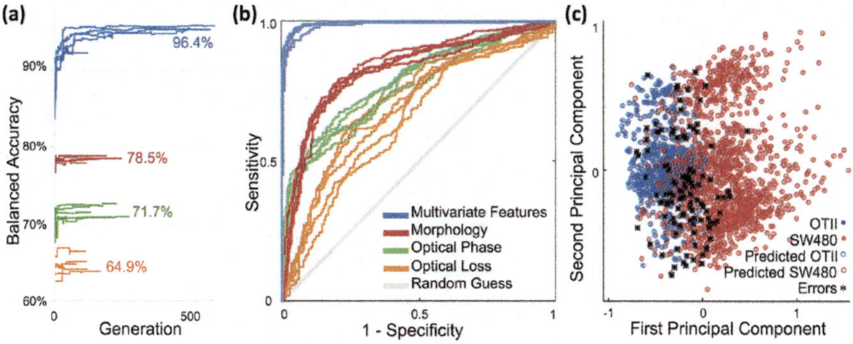

Fig. 8.2 Classification of white blood cells (*OT-II*) and cancer cells (*SW-480*) by TS-QPI label-free features; (**a**) Training process of the neural network leads to improvement of classification accuracy over generations of genetic algorithm. In addition to multivariate analysis using all 16 biophysical features extracted from the TS-QPI quantitative images (*blue curves*), we also show training process by three single features. *Red, green, and orange* curves represent the best biophysical feature in each category, morphology, optical phase, and optical loss, respectively. The values represent average balanced accuracy among training datasets at the end of optimization. Clearly, the final achievable accuracy by multivariate classification is considerably higher than that of single features. (**b**) For each case, we show 5 ROC curves for different test datasets. The gray diagonal line shows results of random guess classification. Multivariate analysis based on TS-QPI images (*blue curves*) shows significant improvement in classification sensitivity and specificity. The fact that the classifiers remain almost unchanged during the five iterations of cross validation shows consistency and robustness of the classifiers. (**c**) To visualize the multivariate classification results, data points are depicted in the space of the first two PCA components

generated based on the test subsets (Fig. 8.2b) and reveal the superior sensitivity and specificity of multivariate classifier. Also, the small variations of the ROC curves among different folds show the consistency of the classification performance for different test datasets. To visualize the hyperspace decision boundary, *OT-II* and *SW-480* data points are shown in first and second principal components analysis (PCA) components (Fig. 8.2c).

8.3.2 *Biofuel: Demonstration in Algae Lipid Content Classification*

Microalgae are considered one of the most promising feedstock for biofuels [141]. The productivity of these photosynthetic microorganisms in converting carbon dioxide into useful carbon-rich lipids greatly exceeds that of agricultural crops. Worldwide, research and demonstration programs are being carried out to develop the technology needed to expand algal lipid production as a major industrial process. Selecting high-yield microalgae with fast growth factors are essential in biofuel production industry. Because algae differ greatly in size and structure, cell size alone

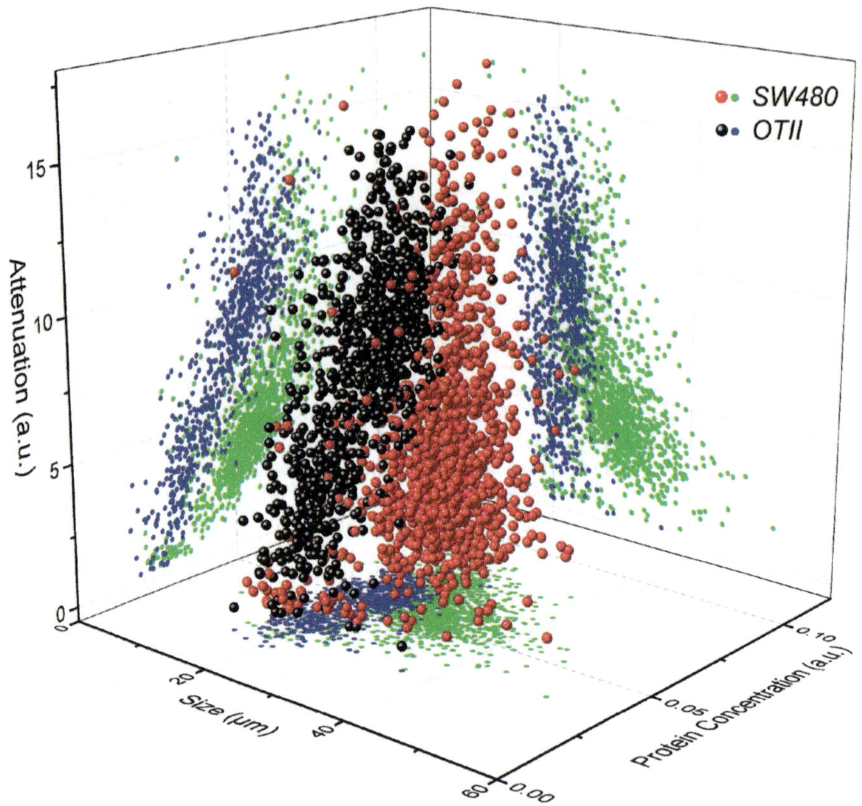

Fig. 8.3 Three-dimensional scatter plot based on size, protein concentration, and attenuation of *OT-II* and *SW-480* cells measured by TS-QPI. The *green and blue dots* are two-dimensional (2-D) projections of cell data points on the planes containing only two of the biophysical features. The cell protein concentration corresponds to the mean refractive index difference of the cell (Refractive index feature in Table 6.2). The attenuation is a feature describing the optical intensity loss caused by cell absorption (Absorption-1 feature in Table 6.2). Comparison of 2-D scatter plots reveals that additional biophysical features (in this case mainly protein concentration) serve to classify the cell types more accurately

provides insufficient information for cell classification. Here we show that adding optical phase and loss data, obtained by the phase contrast time stretch imaging flow cytometer, to size data enables algal cells to be distinguished on the basis of lipid content.

To test our apparatus for its ability to separate algal cells with high and low-lipid content, we exploited the starch-null *sta6* strain of Chlamydomonas. This strain is deleted for *sta6* [142] (encoding the small subunit of ADP-glucose-pyrophosphorylase), and when nitrogen-deprived accumulates more lipid than wild-type [143–146]. *Chlamydomonas reinhardtii* strains used were *cw15* (*nit1 NIT2 mt*$^{+-}$) and *sta6* (*cw15 nit1 NIT2 arg7-7 sta6-1::ARG7 mt*$^{+}$), available as

CC-4568, CC-4348 respectively from the Chlamydomonas resource center (CRC) [147]. Cells were grown in tris-acetate-phosphate (TAP) medium supplemented with arginine ($100\,\mu g\,mL^{-1}$). Cultures were grown in Innova incubators (New Brunswick Scientific, Edison, NJ) at 24 °C, agitated at 180 rpm with continuous light ($95\,\mu mol\,m^{-2}\,s^{-1}$, 6 cool white fluorescent bulbs at 4100 K and 3 warm white fluorescent bulbs at 3000 K per incubator). To induce lipid production, cells were cultured to mid-log phase in regular TAP prior to deprivation of N by transfer to ammonium-free (i.e., nitrogen-free) TAP medium, as described previously [146]. Briefly, cells subjected to nitrogen deprivation were grown to 4×10^6 cells mL^{-1} and collected by centrifugation at 1006 ×g for 5 min at room temperature. The supernatant was discarded, and the cells were washed in nitrogen-free TAP. Cells were then resuspended in nitrogen-free TAP to a final cell count of 2×10^6 cells mL^{-1}. Cell densities were determined using a hemocytometer.

Comparison of the two strains therefore provides an ideal setup to test our ability to distinguish lipid-content phenotypes. Figure 8.4a shows the 3D scatter plot showing the three principle physical features for the two algae populations. Here, the optical loss category of the features plays a dominant role in label-free classification. In Fig. 8.4b, we show ROC curves for binary classification of these

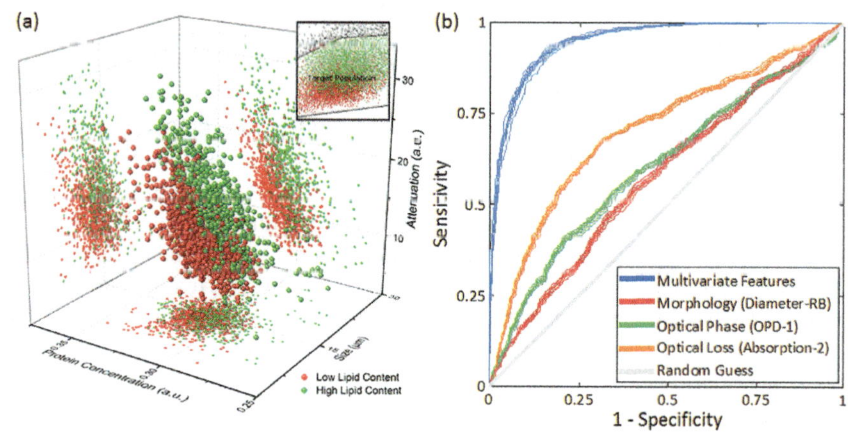

Fig. 8.4 Classification of algal cells (*Chlamydomonas reinhardtii*) based on their lipid content by TS-QPI. (**a**) Three-dimensional scatter plot based on size, protein concentration, and attenuation of the cells measured by TS-QPI, with 2D projections for every combination of two features. Inset: Conventional label-free flow cytometry using forward scattering and side scattering is not enough to distinguish the difference between high-lipid content and low-lipid content algal cells. TS-QPI is much more effective in separating the two algae populations. (**b**) ROC curves for binary classification of normal and lipid-rich algae species using ten-fold cross validation; *blue curves* show the classifier performance using all 16 biophysical features extracted from the TS-QPI quantitative images. *Red, green, and orange curves* show the classifier decision performance using only the best biophysical feature in each category: morphology (Diameter-RB in Table 6.2), optical phase (OPD-1 in Table 6.2), and optical loss (Absorption-2 in Table 6.2). The label-free selection of algal strains improves as more biophysical features are employed

populations. Blue curves show the classifier performance using all 16 physical features extracted from the TS-QPI images. Red, green, and orange curves show the classifier decision made using only the three major biophysical features: diameter for morphology (Diameter-RB in Table 6.2), optical path length difference for optical phase (OPD-1 in Table 6.2), and absorption for optical loss (Absorption-2 in Table 6.2). Our multivariate deep learning using TS-QPI is far more accurate than individual biophysical characteristics for selection of algal strains.

8.4 Further Discussions in Machine Learning

8.4.1 Learning Curves

To show the effect of the training dataset size in the performance of the learning model, the learning curves for the training and test datasets of the tumor cell detection are analyzed (Fig. 8.5a). The test learning curve shows that as the number of training data points increases, the test error reduces and the model performance improves. On the other hand, the training error contrastingly increases for a larger number of training examples because it is more difficult for the learning model to fit many training data points than a few. The discrepancy of the training and test errors is the generalization error of the learning model [128]. Notice that beyond $N \cong 850$ the generalization error do not decrease, and the learning curves converge to their ultimate performances. In other words, $N \cong 850$ training data points are required to accomplish target achievable performance for the deep learning model used here.

Multiple machine learning techniques [129, 134, 148, 149] for multivariate label-free cell classification are compared using our TS-QPI tumor cell detection dataset (Fig. 8.5b). The mean accuracies of all learning models are beyond 85%, reflecting the advantages of simultaneous hyperdimensional biophysical features that TS-QPI provides for label-free cell classification. Furthermore, the interquartile range of the balanced accuracy (shown with box plot) is the smallest for the regularized AUC-based deep learning model, which confirms its consistency and repeatability are the best among learning methods.

8.4.2 Principal Component Analysis (PCA)

As shown in Fig. 6.6a and b, many of the 16 features are correlated and not all measured features in the data set produced by the time stretch quantitative phase imaging have the same amount of information. That result suggests that it may be possible to reduce the 16-dimensional data set to a smaller set of uncorrelated orthogonal dimensions without significantly compromising the classification accuracy. In that spirit, we have used principle component analysis (PCA) for

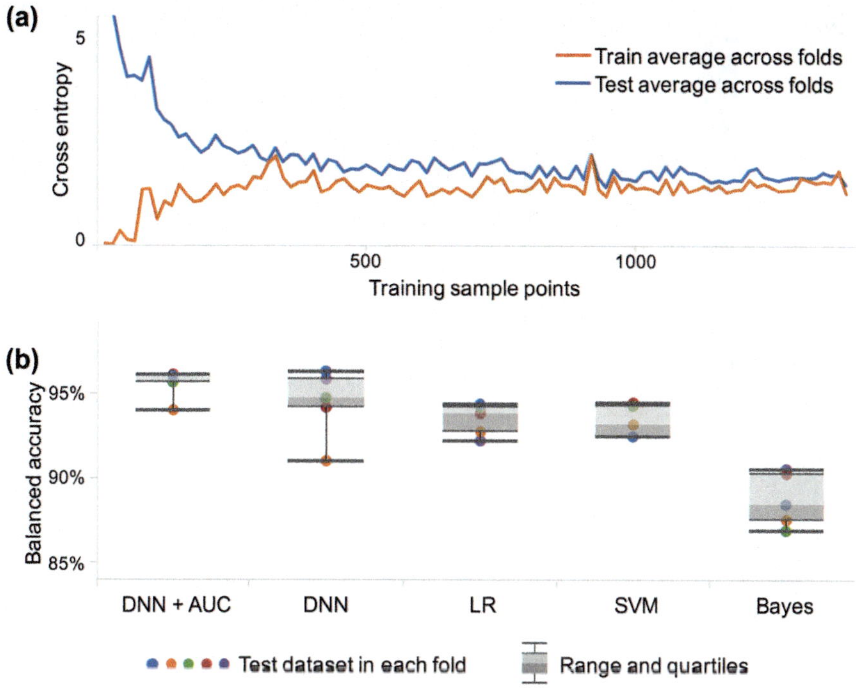

Fig. 8.5 (**a**) The learning curves of the training and test datasets in the tumor cell detection. Larger number of training data points decreases the cross entropy of the test dataset, which means the classifier is performing more accurately. However, the trend is opposite for the training dataset because with a larger number of training data points fitting error accumulates. The discrepancy of the training and test errors, i.e., generalization error, decreases up to $N \cong 850$, which is the necessary training data size for achieving final performance in our TS-QPI demonstration with deep learning neural network. (**b**) Comparison of multiple machine learning classification techniques based on the biophysical features extracted from the label-free cell images captured by TS-QPI. Our AUC-based deep learning model (DNN+AUC) has both the highest accuracy and consistency against support vector machine (SVM) with Gaussian kernel, logistic regression (LR), naive Bayes, and conventional deep neural network trained by cross entropy and backpropagation (DNN)

dimensionality reduction and computation speed-up. The PCA algorithm finds an alternative lower dimension space such that variance of data projected onto this subspace is maximized along subspace dimensions. Figure 8.6a shows the percent of the variance in data explained by each component (lower chart). The key observation is that most of the variance can be accounted for by the first two principle components. The upper portion of the plot shows the accuracy for binary classification using each of the principle components. Interestingly, the first component with the highest explained variance is not necessarily the most important component for classification. Therefore, a priori intuition about the physical significance of the features in the case here is superior to PCA in

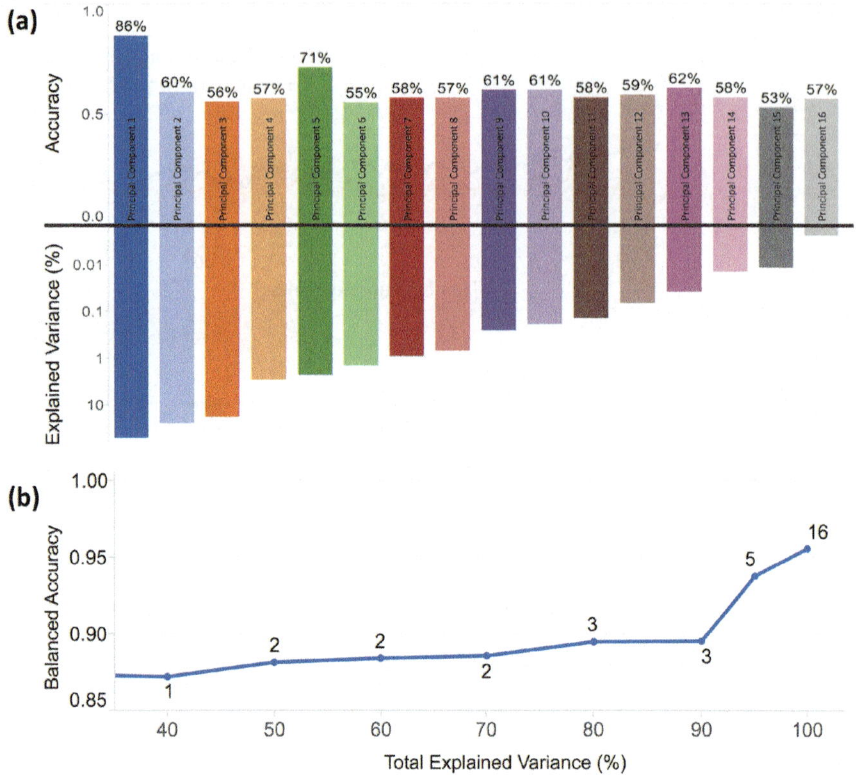

Fig. 8.6 Principal component analysis (PCA) on the multivariate data set produced by time stretch quantitative phase imaging. (**a**) Upper bar chart shows accuracy of classification by each individual principal component, and lower bar chart shows the percentage of the total variance explained by each principal component, accounting for the variability expressed in the data. As expected, principal components with larger variability do not necessarily give high accuracy in classification. (**b**) Cumulative accuracy. The value at each data point corresponds to the number of PCA components retained in order to achieve that total explained variance. In order to reduce the number of input features and decrease computation time, a subset of the PCA components can be used for classification. The classification accuracy improves as the total variance retained in the subset of PCA components goes up

eliminating dimensions that do not provide high value in classification. By revealing the structure in data that best explains the variance, PCA achieves data compression via dimensionality reduction.

PCA components act as the input features for the classification algorithm. As number of PCA components retained increases, the classification accuracy improves while computation time increases (Fig. 8.6b). The value at each data point corresponds to the number of PCA components retained in order to achieve that total explained variance. In order to reduce the number of input features and decrease computation time, a subset of the PCA components can be used for classification.

The classification accuracy improves as the total variance retained in the subset of PCA components goes up. Nearly 90% accuracy can be achieved with the first three PCA components. The small deviation among accuracies of data points with the same number of PCA components is due to variations in random data partitioning. Since accuracy is the main concern here, we employ all 16 biophysical features, rather than dimensionality-reduced PCA components.

8.4.3 Cross Validation

The k-fold cross validation implemented here splits data points into training, validation, and test subsets (Fig. 8.7a). For each iteration, one fold is used as test data, one for validation, while the other folds are used during training process. After initially trained, the performance of the network is analyzed by the validation data to fine tune the neural network architecture and regularization parameter. Figure 8.7b shows that either a too small or a too large regularization parameter, λ, increases network error due to overfitting or underfitting, respectively. Therefore, there is a suitable range of regularization parameter for each learning model.

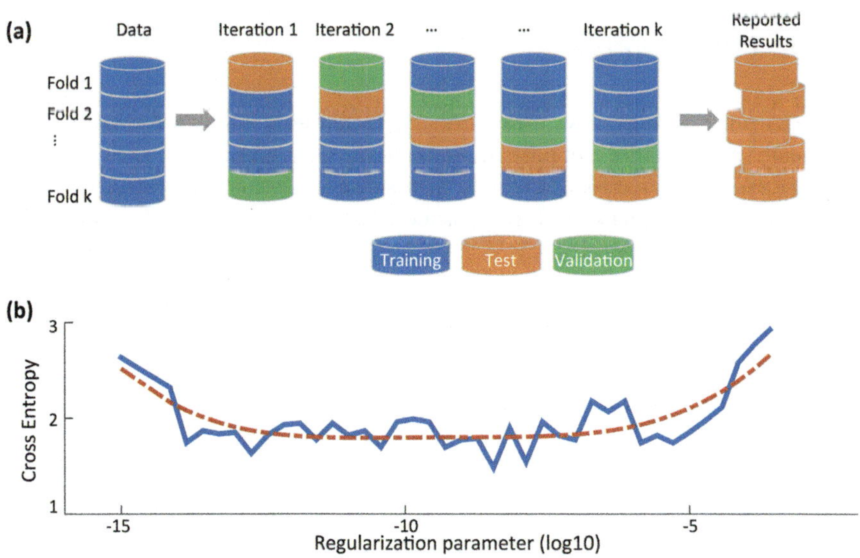

Fig. 8.7 (**a**) The implementation of the k-fold cross-validation here splits data points into training, validation, and test subsets. In each iteration, one fold is used for fine tuning the learning model (validation dataset) and another fold is used for evaluation of the final results (test dataset), while rest of the data points are used for training (training dataset). The final reported results are aggregate of the outcomes from the test datasets. (**b**) A suitable regularization parameter, λ, balances the trade-off between overfitting (variance) and underfitting (bias) and minimizes the cross entropy of the validation dataset

Once the network architecture and regularization parameter are chosen and optimized based on the validation data, the learning model performance is finally verified by the test fold, which has never been used before in this iteration. The process of training in each iterations is independent, so each iteration has no prior knowledge about the chosen learning models in other iterations. The final reported results are aggregate of the performance for different test datasets.

8.4.4 Computation Time

Our deep learning technique uses AUC as the cost function and performs training via genetic algorithm. Since AUC is calculated based on the entire dataset, the genetic algorithm is employed as a global optimization method [150]. Thus, our technique has inherently higher accuracy and repeatability compared to conventional deep learning and other classification algorithms studied here. However, the global optimization in our algorithm sacrifices the computation time. The performance of balanced accuracy and computation time of different classification algorithms are compared in Table 8.1.

8.4.5 Data Cleaning

Data cleaning includes two steps. Firstly, Hotelling's T-squared distribution is calculated and the top 2% of the extreme data was set as outliers due to experimental or object recognition errors. Secondly, debris discrimination is performed; any data point with negative phase shift was considered as either air bubble, flow turbulence, or object recognition errors.

Table 8.1 Performance comparison of different classification algorithms

Algorithm	Averaged balanced accuracy (%)	Standard deviation of balanced accuracy (%)	Computation time (s)
Deep neural network trained by AUC	95.5	0.9	365.6
Deep neural network trained by cross entropy	94.4	2.1	4.7
Logistic regression	93.5	0.9	0.8
Support vector machine	93.4	1.0	1.7
Naive Bayes	88.7	1.6	2.8

8.5 Conclusion

Time stretch quantitative phase imaging (TS-QPI) is capable of capturing images of flowing cells with minimal motion distortion at unprecedented rates of 100,000 cells/s. TS-QPI relies on spectral multiplexing to capture simultaneously both phase and intensity quantitative images in a single measurement, generating a wealth of information of each individual cell eliminating the need for labeling with undesirable biomarkers. Here, we summarized the information content of these images in a set of 16 features for each cell, and performed classification in the hyperdimensional space composed of these features. We demonstrated application of various learning algorithms including deep neural networks, logistic regression, naive Bayes, as well as a new training method based on area under the ROC curve. The results from two experimental demonstrations, one on detection of cancerous cells among white blood cells, and another one on identification of lipid-rich algae, show that classification accuracy by using the TS-QPI hyperdimensional space is more than 17% better than the conventional size-based techniques. Our system paves the way to cellular phenotypic analysis as well as data-driven diagnostics, and thus is a valuable tool for high-throughput label-free cell screening in medical, biotechnological, and research applications.

Part V
Data Compression

Chapter 9
Optical Data Compression in Time Stretch Imaging

Time stretch imaging offers real-time image acquisition at millions of frames per second and subnanosecond shutter speed, and has enabled detection of rare cancer cells in blood with record throughput and specificity. An unintended consequence of high-throughput image acquisition is the massive amount of digital data generated by the instrument. Here we report the first experimental demonstration of real-time optical image compression applied to time stretch imaging. By exploiting the sparsity of the image, we reduce the number of samples and the amount of data generated by the time stretch camera in our proof-of-concept experiments by about three times. Optical data compression addresses the big data predicament in such systems.

9.1 Background

Big data is a broad and popular topic today. The traditional definition refers to the massive amount of data generated in banking, social media, healthcare, and by networked sensors known as the "internet of things." However, big data is also a challenge in biomedical and scientific instruments [22]. High-throughput real-time instruments are needed to acquire large data sets and to detect and classify rare events. Examples include the time stretch camera [10–12, 15–17, 19, 20, 46, 151, 152]—a MHz-frame-rate bright-field imager, and the fluorescence imaging using radio frequency-tagged excitation (FIRE)—an ultra-high-frame-rate fluorescent camera for biological imaging [21]. The record throughputs of these instruments have enabled the discovery of optical rogue waves [7], the detection of cancer cells in blood with false positive rate of one cell in a million [30], and the highest performance analog-to-digital converter ever reported [8]. These instruments produce a torrent of data that overwhelms their data acquisition and processing backend. For example, the time stretch imager captures images at

A. Mahjoubfar et al., *Artificial Intelligence in Label-free Microscopy*,
DOI 10.1007/978-3-319-51448-2_9

roughly one hundred million scans per second with each scan containing about one thousand samples [13, 14]. Assuming each of these samples is digitized with a typical 8 bits of accuracy, time stretch microscopy (STEAM) produces 0.8 Tbit of data per second. Detecting rare events such as cancer cells or rogue signals requires that data be recorded continuously and for a long time to catch the rare events. The need to compress massive volumes of data in real-time has fueled interest in nonuniform time stretch transformation that takes advantage of sparsity in physical signals to achieve both bandwidth compression and reduction in the temporal length [22–26]. The aim of this technique is to transform a signal such that its intensity matches not only the digitizer's bandwidth, but also its temporal record length. The latter is typically limited by the digitizer's storage capacity.

9.2 Warped Stretch Imaging

The basic principle of time stretch imaging (STEAM) involves two steps both performed optically. In the first step, the spectrum of a broadband optical pulse is converted by a spatial disperser into a rainbow that illuminates the target. Therefore, the spatial information (image) of the object is encoded into the spectrum of the resultant reflected or transmitted rainbow pulse. A one-dimensional rainbow is used to acquire a line-scan and two-dimensional image is obtained by scanning the rainbow in the second dimension. For imaging of particles in flow, the motion causes scanning in the second dimension while the rainbow position is fixed (Fig. 9.1).

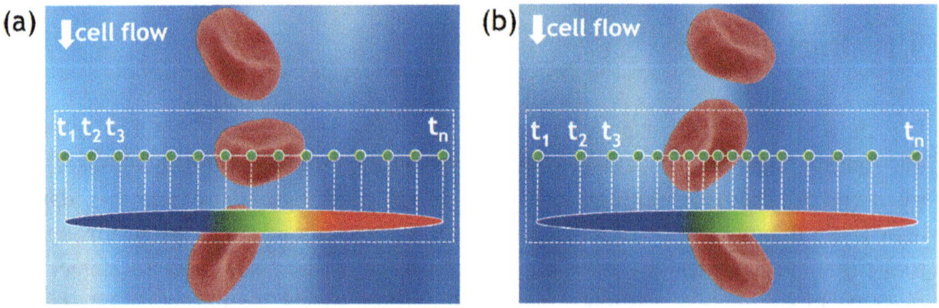

Fig. 9.1 Illustration of warped stretch transform in imaging. (**a**) The field of view consists of a cell against the background such as a flow channel or a microscope slide. Illumination by an optical pulse that is diffracted into a one-dimensional rainbow maps one dimension of the space into the optical spectrum. The other dimension is scanned by the cell flow through the rainbow. In the conventional time stretch imaging (STEAM), the spectrum is linearly mapped into time using a dispersive optical fiber with a linear group delay. The temporal waveform is then sampled by a digitizer with fixed sampling rate resulting in uniform spatial sampling. But uniform spatial sampling generates superfluous data by oversampling the sparse peripheral sections of the field of view. (**b**) Similar functionality can be achieved in STEAM by using a nonlinear group delay profile in the spectrum-to-time mapping process resulting in a nonuniform sampling of the line image, assigning more pixels to the information-rich central part of the field of view and less to the low-entropy peripherals

In the second step, the spectrum of the image-encoded pulse is mapped into a serial temporal signal that is stretched in time to slow it down such that it can be digitized in real-time [9]. This optically amplified time stretched serial stream is detected by a single-pixel photodetector, and the image is reconstructed in the digital domain. Subsequent pulses capture repetitive frames. The laser pulse repetition rate corresponds to the frame rate, and the temporal width of the pulses corresponds to camera's shutter speed (exposure time). The key innovations in STEAM that enable high-speed real-time imaging are photonic time stretch for digitizing fast images in real-time and optical image amplification for compensating the low number of photons collected during the ultra-short shutter time [29].

Using warped group delay dispersion, it has been shown that one can reshape the spectro-temporal profile of optical signals such that signal envelope's time-bandwidth product is compressed [22–26]. The compression is achieved through time stretch dispersive Fourier transform in which the frequency-to-time mapping is intentionally warped, using an engineered group delay dispersion profile, to match the sparsity of the image. This operation causes a frequency-dependent reshaping of the input waveform. Reconstruction (decoding) method depends on whether the information is in the spectral domain amplitude, or in the complex spectrum. In the time stretch camera, the image is encoded into the amplitude of the spectrum of a broadband optical pulse, and reconstruction consists of a time-to-frequency mapping using the inverse of the measured or simulated group delay profile followed by a frequency-to-space mapping. The compression ratio depends on the group delay characteristics and the sparsity of the image [25, 26]. This method offers similar functionality as compressive sampling [18, 27, 28, 153–156] albeit it achieves it via an entirely different approach, namely by reshaping the analog image using warped time stretch dispersive Fourier transform.

9.3 Optical Image Compression

To illustrate the concept in the context of time stretch imaging, we can consider a microscopic field of view consisting of a cell against a background such as a flow channel or a microscope slide (Fig. 9.1a, b). In the time stretch imaging, the object is illuminated by an optical pulse that is diffracted into a one-dimensional rainbow. This encodes one dimension of space into the optical spectrum. The spectrum is then linearly mapped into time using a dispersive optical fiber with a linear group delay. The mapping process from space-to-frequency-to-time is shown in Fig. 9.1a. The linearly stretched temporal waveform is then sampled by a digitizer resulting in uniform spatial sampling. This uniform sampling (also depicted in Fig. 9.1a) generates redundant data by oversampling the sparse peripheral sections of the field of view.

Such a situation evokes comparison to the mammalian eye where central vision requires high resolution while coarse resolution can be tolerated in the peripheral vision (Fig. 9.2a). In the eye, this problem is solved through nonuniform

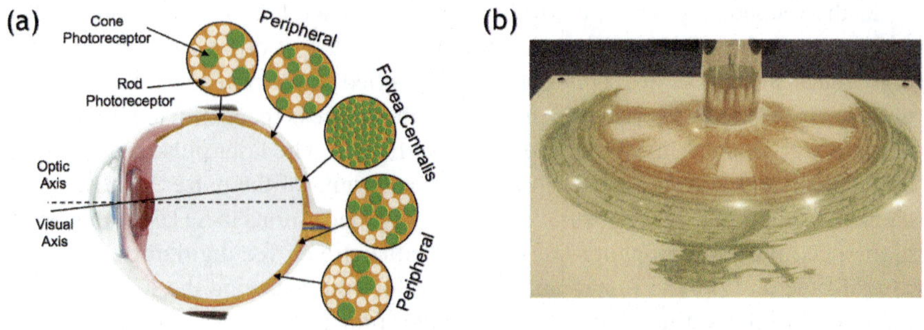

Fig. 9.2 Warped stretch transform in imaging inspired by biology and art. (**a**) The human vision is a form of warped imaging system where high sampling resolution is needed in the central vision while coarse resolution can be tolerated in the peripheral vision. (**b**) The reconstruction is similar to anamorphic art where the drawn shape is a stretched and warped version of the true object yet the viewer sees the true object upon reflection from a curved mirror. In our system, this unwarping operation is a nonlinear mapping using the inverse space-to-frequency-to-time mapping transfer function

photoreceptor density in the retina. The Fovea section of the retina has a much higher density of photoreceptors than the rest of the retina and is responsible for the high resolution of central vision.

We solve this problem by nonuniform mapping of spectrum into time via a warped group delay. An example of the warped space-to-frequency-to-time mapping is illustrated in the dotted box in Fig. 9.1b. After uniform sampling in time (by a conventional digitizer), this leads to higher sampling density in the central field of view and lower density in the sparse peripheral regions. This is often desirable in cell screening and imaging in microfluidic channels with focusing mechanisms. In these channels, the cells arrive along a few predetermined lanes. By far the most common case is a single lane aligned to the center of the channel, which is typically achieved via hydrodynamic focusing [80]. However, cells or particles may occasionally appear in peripheral regions of the flow channel. Since the probability of this occurring is low, it would be wasteful to assign high sample density to these peripheral regions. One does need to image these regions albeit with coarse resolution for monitoring rare or abnormal events. In the meantime, the higher sample density in central part of the field of view improves the accuracy of determining cellular morphology, and that of biophysical cell measurements such as cellular protein concentration, which have been previously demonstrated with the time stretch imaging modality [17, 19].

The reconstruction is a simple unwarping using the inverse of the group delay. This operation is analogous to the anamorphic art, where the drawn shape is a stretched and warped version of the true object, yet, the viewer sees the true object upon reflection of the painting from a curved mirror (Fig. 9.2b [157]). In the case where the sparsity characteristic of the target is not known, or changes dynamically, a shift of central field of view is needed. Similar to the movement of the eyeball in the mammalian eye, an active mechanism such as a beam steering mirror can be used to relocate the central field of view, and perform the dense sampling in the region of interest.

Different nonlinear group delay profiles result in various types of warped frequency to time mappings. The linear group delay profile results in uniform frequency-to-time mapping (orange plots), whereas the warped group delay profile results in nonuniform mapping (blue plots). In both the linear and the warped stretch cases, the stretched waveform is uniformly sampled in time and digitized by an analog-to-digital converter. The amount of linear dispersion is chosen such that the fast features of the waveform are sufficiently slowed down in time to achieve Nyquist sampling. Figure 9.3a shows a nonuniform group delay profile, which has the same dispersion (slope) as the linear profile in the center of the spectrum, but reduced dispersion at wings. This profile results in data compression by reduction of the overall time duration of the stretched pulses and the number of samples at the expense of lowered spectral resolution in peripheral regions of the spectrum. Here the digital file size, determined by the overall number of samples, is significantly reduced (compare the waveforms). In other words, data is compressed in optical domain by exploiting its sparsity. Figure 9.3b shows another nonuniform group delay profile, which has the same overall time duration and number of samples as the linear case. This profile redistributes the spectral samples to achieve higher spectral resolution in information-rich central region of the spectrum and lower resolution in sparse peripherals. Here the file size determined by the overall number of samples in the waveform and the acquisition time remains unchanged.

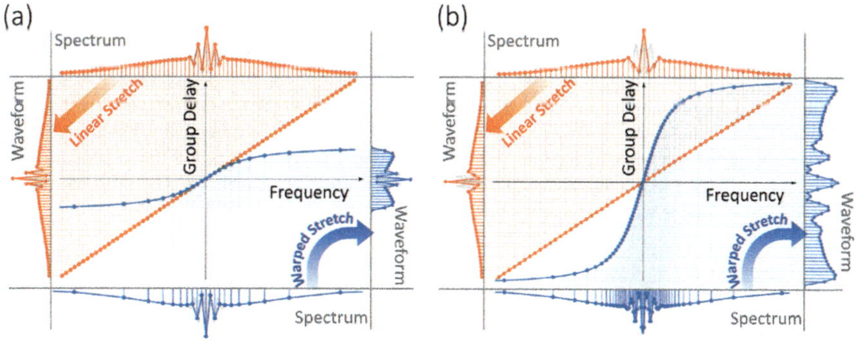

Fig. 9.3 Linear and warped (anamorphic) stretch transforms. The linear group delay profile results in uniform frequency-to-time mapping (*orange plots*), whereas the warped group delay profile results in nonuniform mapping (*blue plots*). (**a**) A nonlinear group delay with the same dispersion (slope) at the center of the spectrum as linear case, but shorter total group delay, leads to high sampling resolution in the center of the spectrum and lower resolution at the wings. This keeps the image quality at the central part of the field of view intact, while reducing the quality at the sparse peripheral regions where uniform stretch would have produced redundant samples. (**b**) A nonlinear group delay profile with higher dispersion (slope) at the center of the spectrum than the linear case, but same total group delay over the bandwidth, leads to a higher spectral resolution in the center of the spectrum and lower resolution at the wings (compare the spectrums). The *gray curves* show the analog waveforms before sampling for the purpose of comparison

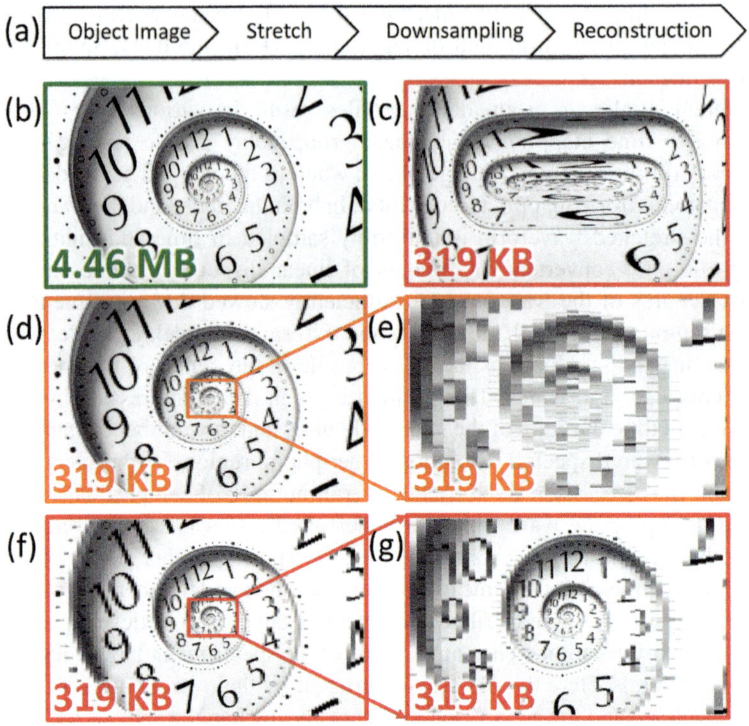

Fig. 9.4 Simulations illustrate the effect of warped stretch transform on a two-dimensional image. The analog reshaping of the image performed in the optical domain by the warped stretch transform is emulated here in the digital domain. (**a**) The transformation consists of nonuniform stretch in the horizontal direction with the warp stretch profile shown in Fig. 9.3. (**b**) A sample image with 28,001,672 pixels and 4.46 MB file size is used as the input. (**c**) The image is stretched nonuniformly and down-sampled with a compression ratio of 14. (**d**) A uniform stretch with down-sampling can achieve the same file size but the image quality is dramatically lower. (**e**) While down-sampling is not an issue for the sparse peripherals, it is problematic for the information-rich central part. (**f**) The reconstruction of the nonuniformly stretched image. (**g**) The information-rich region at the center is well preserved while maintaining the same sampling rates

To help visualize the analog image reshaping performed by warped dispersive stretch and to show how it leads to data compression in imaging, we emulate its effect on a two-dimensional image. As shown in Fig. 9.4a, the image is first stretched and then uniformly down-sampled to achieve data compression, followed by reconstruction (unstretch). By using a nonlinear stretch, the reconstructed image is equivalent to a nonuniformly down-sampled image. Figure 9.4b shows the original image as if it was generated by a linear dispersion and uniform stretch. It corresponds to an image formed by a linear stretch followed by Nyquist rate sampling. Figure 9.4c is the result of nonuniform stretching of the original image in the horizontal direction. The chosen image has higher density of features in the central portion than in the periphery. The warp profile is as indicated in Fig. 9.3b where the peripheral regions are stretched less than the center. Figure 9.4d is the linearly stretched image after 14:1 down-sampling and reconstruction. The warped stretch profile used here is as shown in Fig. 9.3b causing the sides to be squeezed relative to the center. Hence more samples are assigned to the center than the wings. As it can be seen in the zoomed-in Fig. 9.4e, down-sampling has resulted in a loss of resolution. On the other hand, Fig. 9.4f is the nonuniformly stretched image after 14:1 down-sampling followed by reconstruction. The resolution is higher at the center where information is rich and lower at the sides where information is much less and relatively not important. Although the final image size is the same, the nonuniformly stretched image has much higher quality in the non-sparse center of field of view (Fig. 9.4g).

Big data problems also appear in light scattering based flow cytometry. There the instrument measures the angular dependence of laser light scattered by particles in flow. The angular scattering profile of microscopic particles significantly depends on their morphological parameters, such as size and shape, and this dependency is widely used in flow cytometry for particle classification [95]. Recently a new spectrally encoded angular light scattering method capable of measuring the continuous angular spectrum has been reported [158]. The warped time stretch optical data compression technique demonstrated here can also be used for real-time data compression in such optical systems.

9.4 Experimental Design and Results

The experimental setup used for our proof-of-principle demonstration of optical image compression is shown in Fig. 9.5. A mode-locked fiber laser generated pulses at around 1550 nm with a repetition rate of 36.129 MHz and a pulse width slightly less than 100 fs. A short dispersion compensating fiber with an overall dispersion of 10 ps/nm was used to temporally broaden pulses to about 1 ns, so that an erbium-doped fiber amplifier (EDFA) can amplify them without any distortion. Since the output spectrum of EDFA is sensitive to the input polarization, a polarization controller was used to change the polarization of the input pulses to EDFA. The polarization was tuned in such a way that the output amplified pulses had relatively symmetric spectrum around 1550 nm. Amplified pulses then entered a coarse wavelength-division multiplexing (WDM) filter, and the output of 1551 nm channel was used to shape laser pulses with a considerably low noise floor over 1541–1561 nm bandwidth. These pulses passed through an optical circulator and were coupled to free-space with a fiber collimator.

Free-space laser pulses were linearly polarized with quarter- and half-wave plates, and then they were spatially dispersed with a pair of reflection diffraction gratings, so that each wavelength component of the collimated beam was positioned at a different lateral point similar to a rainbow. The width of the rainbow depended on the size of the second diffraction grating and the distance between two diffraction gratings, and the height of the rainbow depended on the beam size from the fiber

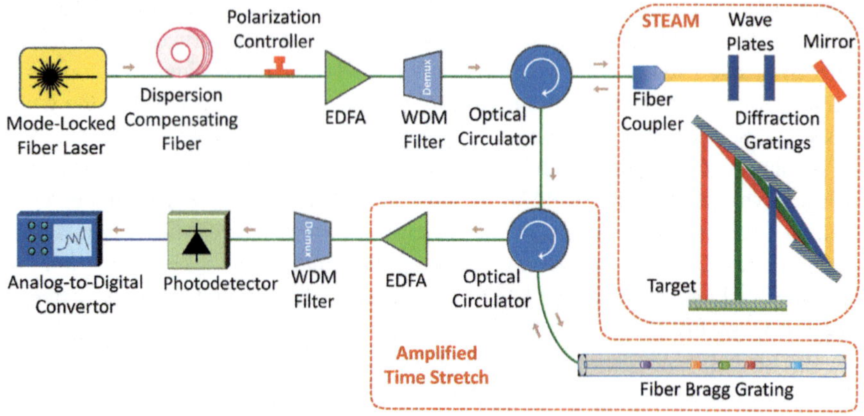

Fig. 9.5 Experimental setup used in proof-of-concept demonstration of optical data compression via warped time stretch imaging. A train of broadband optical pulses was generated at 1550 nm central wavelength with a repetition rate of 36.129 MHz and a pulse width slightly less than 100 fs. The laser pulses were temporally stretched to about 1 ns by a dispersion compensating fiber and amplified by an erbium-doped fiber amplifier (EDFA). The bandwidth over 1541–1561 nm was selected by a wavelength division multiplexing (WDM) filter. The pulses passed through an optical circulator and were coupled to free-space part of STEAM setup with a fiber collimator. There, a pair of diffraction gratings generates a one-dimensional rainbow with each wavelength component imaging a different location at the target. The spectrally encoded rainbows are reflected and coupled back into the fiber, carrying the image information. The nonuniform space-to-frequency-to-time mapping is achieved with a warped chirped fiber Bragg grating (CFBG). After optical image amplification by another EDFA, different wavelength components are detected serially by a single-pixel photodetector and acquired by an analog-to-digital converter (ADC)

collimator. In this setup, the total bandwidth of the pulses interrogating the target is limited to about 10 nm centered at 1551 nm because of the clipping of the rainbow at the edges of the second diffraction grating. The horizontal field of view, which was dictated by the width of the rainbow, was 5 cm. Different wavelength components of the rainbow reached a reflective object. Each pulse of the mode-locked laser generates one rainbow, which captures one line image across the field of view. The rainbow components located at the target were reflected back (Fig. 9.6a) and returned all the way back to the fiber, where they were directed with the optical circulator to an amplified time stretch system (Fig. 9.5). The nonlinearly dispersed pulses with chirped group delay profile are captured by a 10 GHz-bandwidth single-pixel photodetector and digitized in real-time. An analog-to-digital convertor (ADC) with a sampling rate of 20 GSps and 7 GHz bandwidth was used to digitize the output signal of the photodetector. To achieve warped stretch, we used a fiber Bragg grating with customized chirp profile, whose performance was studied in [24] and is shown in Fig. 9.6b.

The two-dimensional image was reconstructed by stacking spectrally encoded horizontal line images at different steps of the vertical scan. If instead of the fiber Bragg grating, a dispersive fiber with linear group delay was used, the reconstructed

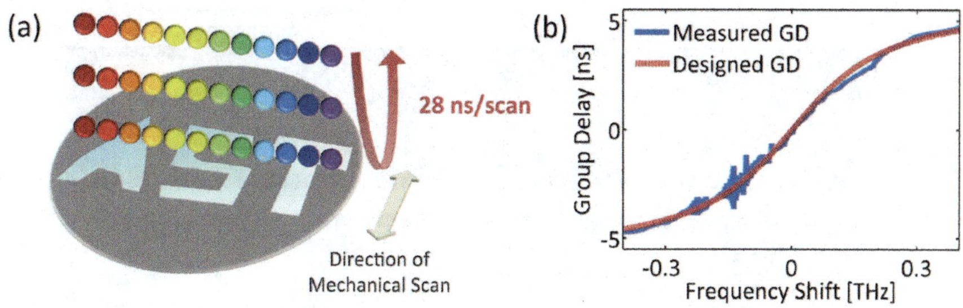

Fig. 9.6 Proof-of-concept experimental setup. (**a**) The test sample reflected one-dimensional rainbow illumination pulses, which are used to perform time stretch imaging at a scan rate of 36 MHz. The field of view determined by the length of the rainbow was 5 cm and covers the width of the target. The vertical direction was scanned by mechanical translation at 0.5 mm per step. (**b**) The warped stretch transform leading to nonlinear spectrum-to-time mapping is performed by a custom chirped fiber Bragg grating with sublinear group delay (GD) profile. This profile gives higher group delay dispersion at the center frequency and reduced dispersion at the wings of the bandwidth

image from one pulse per horizontal line was as shown in Fig. 9.7a. But, for the case of a fiber Bragg grating, since each line-scan is warped, the warping of the image is observed in the horizontal direction (Fig. 9.7b). This effectively means that the central area (letter "*S*") is sampled with higher resolution than the peripherals (letters "*A*" and "*T*"). With the unwarping algorithm derived from the reverse dispersion profile, the uniform image was successfully reconstructed with a reduced data acquisition time and number of samples (Fig. 9.7c). Compared to the case of the linear group delay (Fig. 9.7a), an image with comparable quality is generated with only one-third of the data size (Fig. 9.7c). We note that the reconstruction is an intensity-only operation and does not require optical phase retrieval. Images with improved quality can be generated by averaging many pulses to form each horizontal line image. However, this reduces the frame rate of the time stretch camera. Figure 9.7d–f shows such images formed by averaging 722 pulses for each horizontal line. Although the image quality is slightly better using averaging, but in our demonstration, the signal-to-noise ratio even in single-pulse acquisition mode (Fig. 9.7a–c) is high enough that the target features are clearly recognizable, and there is no need for averaging. This is due to the relatively high pulse-to-pulse stability of the STEAM setup. Warped group delay profiles used here are only a few cases of the unlimited variety of nonlinear space-to-frequency-to-time mappings that can be integrated into time stretch imaging, each corresponding to their unique nonuniform sampling patterns. As another example, Fig. 10.1 shows the nonlinear frequency-to-time mapping profile that is designed for a microfluidic channel with two focal regions (cell flow lanes), a common case in inertial focusing [68, 159]. The profile shown here is designed to have two high-resolution sampling areas corresponding to where the cells are confined. Three low-resolution sampling regions provide coarse resolution in the peripherals regions between the cell flow lanes. The nonuniform mappings can even be reshaped dynamically based on

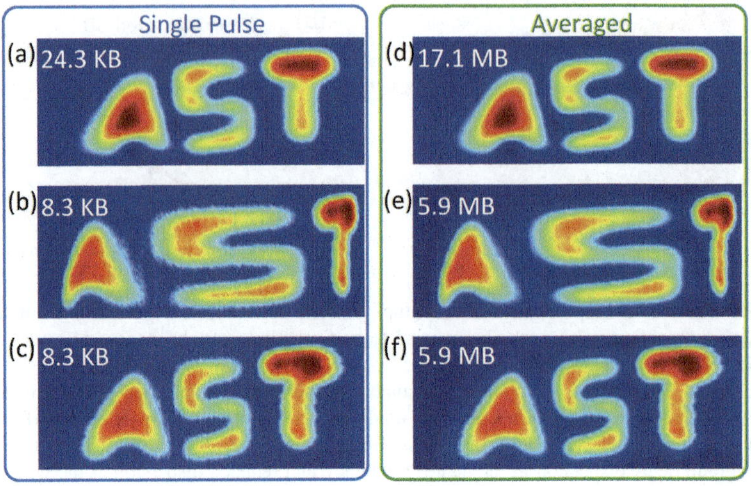

Fig. 9.7 Proof-of-concept experimental results. (**a**) If we use a linear group delay profile with the same dispersion as that of the warped stretch at the center frequency and a single pulse per image line, the image data size would be 24.3 kB (55,345 measured pixels). (**b**) The single-pulse reconstructed image based on the waveform nonlinearly stretched by the chirped fiber Bragg grating has an obvious warping effect at the center of the field of view (letter "S") (18,945 measured pixels). (**c**) The single-pulse unwarped reconstructed image data size is 8.3 kB achieving about three times optical image compression (18,945 measured pixels). (**d**, **e** and **f**) When many pulses (722 pulses here) are averaged to form each horizontal line image, the images' quality improve only slightly over Fig. 9.7a–c, proving high signal-to-noise ratio of our camera even in single-pulse capture mode. The number of measured pixels used in the formation of Fig. 9.7d is 72,255,345, and for Fig. 9.7e, f is 72,218,945. The temporal durations of the waveforms corresponding to each horizontal line in Fig. 9.7a–f are 27.7 ns, 9.5 ns, 9.5 ns, 20.0 μs, 6.8 μs, and 6.8 μs, respectively

relatively slower transitions in the sparsity characteristics of the image, in other words, alterations in the information rich areas of the image. To achieve such a functionality, the group delay profile of the dispersive element should be tunable and controlled by a feedback mechanism. In terms of tunable dispersion, Chromo-Modal Dispersion (CMD) offers wide tunability, broad spectrum, and low loss [160].

Furthermore, warped stretch imaging is not limited to using warped group delay to perform nonlinear space-to-frequency-to-time mapping. It can also be achieved by nonuniform space-to-frequency mapping, e.g., warped rainbows, where rainbow frequency components are not equally spaced. This can be implemented by frequency-dependent spatial dispersers such as custom-designed diffraction gratings and virtually imaged phased arrays.

9.5 Conclusion

Real-time optical image compression is needed to address the fundamental challenges in acquiring and storing the large amount of data generated in high-speed imaging. Here, we have demonstrated one such technique applied to time stretch imaging. Using warped group delay dispersion, we achieved warped stretch imaging in such a way that the information-rich central vision is sampled at a higher sample density than the sparse peripheral vision. Most notably, this was done using a uniform electronic sampler, i.e., without adaptive or dynamic control over the electronic sampling rate. A three-time image compression was achieved in experimental proof of concept demonstration. Our nonuniform sampling technique could offer one route to taming the capture, storage, and transmission bottlenecks associated with big data.

Chapter 10
Design of Warped Stretch Transform

Time stretch dispersive Fourier transform enables real-time spectroscopy at the repetition rate of million scans per second. High-speed real-time instruments ranging from analog-to-digital converters to cameras and single-shot rare-phenomena capture equipment with record performance have been empowered by it. Its warped stretch variant, realized with nonlinear group delay dispersion, offers variable-rate spectral domain sampling, as well as the ability to engineer the time-bandwidth product of the signal's envelope to match that of the data acquisition systems. To be able to reconstruct the signal with low loss, the spectrotemporal distribution of the signal spectrum needs to be sparse. Here, for the first time, we show how to design the kernel of the transform and specifically, the nonlinear group delay profile dictated by the signal sparsity. Such a kernel leads to smart stretching with nonuniform spectral resolution, having direct utility in improvement of data acquisition rate, real-time data compression, and enhancement of ultrafast data capture accuracy. We also discuss the application of warped stretch transform in spectrotemporal analysis of continuous-time signals.

10.1 Overview

Time stretch dispersive Fourier transform [9, 66, 161] addresses the analog-to-digital converter (ADC) bottleneck in real-time acquisition of ultrafast signals. It leads to fast real-time spectral measurements of wideband signals by mapping the signal into a waveform that is slow enough to be digitized in real-time. Combined with temporal or spatial encoding, time stretch dispersive Fourier transform has been used to create instruments that capture extremely fast optical phenomena at high throughput. By doing so, it has led to the discovery of optical rogue waves [7], the creation of a new imaging modality known as the time stretch camera [10], which has enabled detection of cancer cells in blood with record sensitivity [17, 30, 151], a portfolio of

A. Mahjoubfar et al., *Artificial Intelligence in Label-free Microscopy*,
DOI 10.1007/978-3-319-51448-2_10

other fast real-time instruments such as an ultrafast vibrometer [12, 16], and world record performance in analog-to-digital conversion [8, 162]. The key feature that enables fast real-time measurements is not the Fourier transform, but rather the time stretch. For example, direct frequency-to-time mapping can be replaced by phase retrieval [102] or coherent detection after the dispersion [117] followed by back propagation.

Using warped group delay dispersion as a photonic hardware accelerator [103], an optical signal's intensity envelope can be engineered to match the specifications of the data acquisition back-end [24–26]. One can slow down an ultra-fast burst of data, and at the same time, achieve data compression by exploiting sparsity in the original data [104]. Also called anamorphic stretch transform [24, 25], the warped stretch transform performs a nonuniform frequency-to-time mapping followed by a uniform sampler. The combined effect of the transform is that the signal's Fourier spectrum is sampled at a nonuniform rate and resolution. By designing the group delay profile according to the sparsity in the spectrum of the input signal, more samples are allocated to the information-rich portions of the spectrum and fewer to the information-sparse regions where they would be redundant. The only prior information needed is the sparsity of the signal's spectral features, i.e., information about the ensemble of the signal spectrum. No instantaneous feature detection is required as long as the signal's spectral sparsity is within the design range. As a primary application, the utility of this method has been recently demonstrated in real-time optical image compression [104].

In conventional time stretch dispersive Fourier transform, a linear group delay profile is used to impose a nonzero constant group delay dispersion over the full bandwidth of the optical signal as shown in Fig. 10.1 (orange color). This profile as the kernel of the transform generates a linearly increasing frequency-dependent temporal shift across the bandwidth, which maps the optical spectrum into a temporal waveform detectable with a single-pixel photodetector. In other words, temporal dispersion stretches the optical signal in time into its Fourier transform. If the optical signal is a train of ultrafast pulses such as the output of a mode-locked laser, the spectrum of each individual pulse is mapped into a temporal waveform filling the gaps between pulses. To analyze the signal in digital domain, an analog-to-digital converter (ADC) samples the output of the photodetector at a constant rate, which can be interpreted as a uniform Fourier domain sampling of the signal spectrum. It is important to note that the value of each sample of the temporal waveform corresponds to the integral of the optical spectrum over a spectral resolvable window. For a linear group delay profile, the width of this spectral window is fixed, and it does not change over the bandwidth (see orange stripes in Fig. 10.1).

A warped time stretch dispersive Fourier transform is achieved by a nonlinear group delay profile as the kernel of the transform, which imposes a frequency-dependent group delay dispersion onto the spectral components. An example of a nonlinear group delay profile is shown in blue plots of Fig. 10.1. Here, the optical

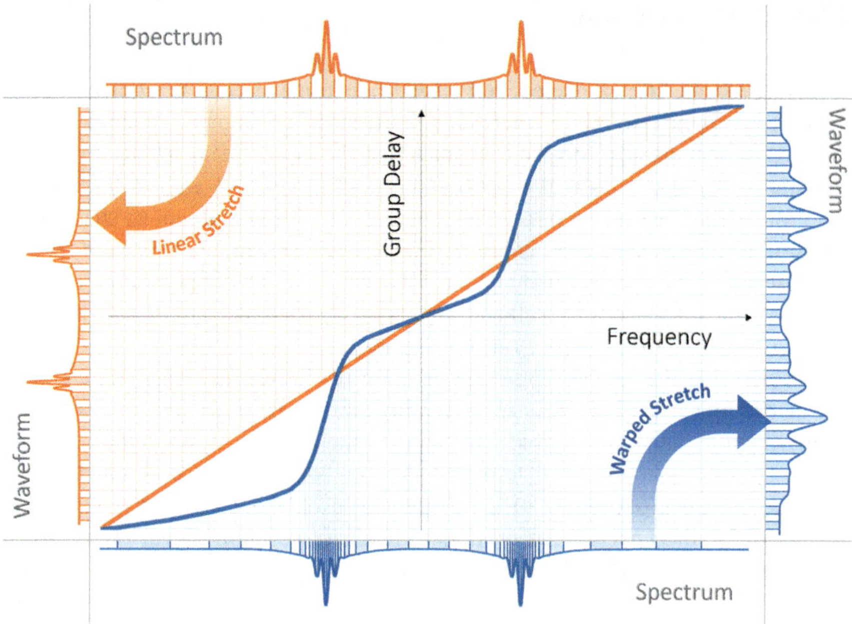

Fig. 10.1 Linear and warped time stretch dispersive Fourier transforms. In linear time stretch (*orange plots*), a linear group delay profile with significant group delay over the signal bandwidth is used as the kernel to delay various spectral components differently, but with a constant group delay dispersion. If the input signal is a pulse train, pulse spectra are linearly mapped to the silent times in between the pulses. In contrast, a nonlinear group delay profile (*blue plots*) with varying group delay dispersion over the bandwidth (different slopes) can stretch the signal spectrum nonlinearly, in which parts of the spectrum are stretched more than the others. This can be used, even with a constant rate sampler, to increase the spectral resolution at regions of the bandwidth where higher resolution is required and to reduce the resolution where the spectral features are sparse (see *blue bars*). In this way, the spectrum is warp stretched into a waveform with the same temporal duration as the linear profile, but smaller bandwidth. In other words, for signals with spectral sparsity, the envelope time bandwidth product can be reduced

spectrum is mapped nonuniformly to time, stretching parts of the spectrum more than the other. In the example shown, the dispersion (slope of the group delay profile) in the central and peripheral parts of the spectrum is smaller than the linear profile, leading to a stretch which is less than the linear profile. However, the profile has also two regions of high dispersion, which map the spectrum into a longer and more-detailed temporal waveform compared to the linear case. It is a remarkable fact that uniform sampling of the warped time stretch output signal corresponds to nonuniform sampling of the spectrum with a frequency-dependent spectral resolvable window (see blue stripes in Fig. 10.1).

10.2 Kernel Design

The warped time stretch dispersive Fourier transformation can be contemplated as a spectrotemporal operation, where its effectiveness in capturing spectral details is dictated by the sparsity and the redundancy of the input signal spectrum. Namely, sparsity in the spectrum is the attribute that influences and guides the design of the group delay dispersion profile. Note that the traditional notion of sparsity, i.e., sparsity in time or spectrum, is not pertinent here. Instead, sparsity here refers to the absence of the spectral features, i.e., abrupt variations of spectrum magnitude. The spectrum only needs to be feature-sparse; it does not need to be narrowband or contain limited number of spectral components.

We describe the spectrotemporal operation of the warped time stretch with a set of examples. Without loss of generality, we initially assume that any chirp in the input signal is negligible compared to the applied group delay dispersion chirp; for example, the total temporal duration of each pulse is much shorter than the overall group delay of the dispersion profile over the pulse bandwidth. In Fig. 10.2a, the envelope of an optical field as the input signal of the stretch transform is shown. The spectrum of this envelope (Fig. 10.2b) has fast variations in the central region (feature-dense) and is relatively smooth in the wings (feature-sparse). The spectrum of the envelope can also be viewed as the spectrum of the input optical field downshifted to baseband. To examine the local properties of the spectrum, we use the short-term Fourier transform of the spectrum and form the spectrogram of the spectrum. This is equivalent to viewing the spectrum as a temporal waveform and plotting its short-time Fourier transform (Fig. 10.2c). As a result, the horizontal axis is the input frequency and the vertical axis is the local frequency of the variations in spectrum magnitude. We call this the frequency of spectrum, which corresponds to the period of variations in the spectrum magnitude or the temporal distance of frequency components of the signal. Here, the window of the short-term Fourier transform is slid over the envelope frequencies for the input envelope spectrum magnitude, and the Fourier transform of the signal in the window gives the local frequencies of spectrum. At the center of the spectrum magnitude, there are fast oscillations (see Fig. 10.2b), which result in high frequencies of spectrum. Hence, the local frequency bandwidth is broad in the central region and narrow in the wings (Fig. 10.2c). A linear group delay profile (Fig. 10.2d) performs conventional time stretch dispersive Fourier transform, in which the spectrum is uniformly mapped into a temporal waveform (Fig. 10.2e). The short-time Fourier transform of this temporal waveform (Fig. 10.2f) resembles that of the spectrum (Fig. 10.2c). To reduce the required acquisition bandwidth or time duration (memory), the spectrotemporal distribution of the signal can be reshaped by a nonlinear group delay profile as a filter, whose characteristics conform to the local frequency patterns. In the regions where the spectrum magnitude has fast variations, the filter should present a high group delay dispersion (slope) resulting in larger stretching in time than the slow varying regions of the spectrum magnitude. For the signal spectrum shown in Fig. 10.2b, a desired group delay profile is shown

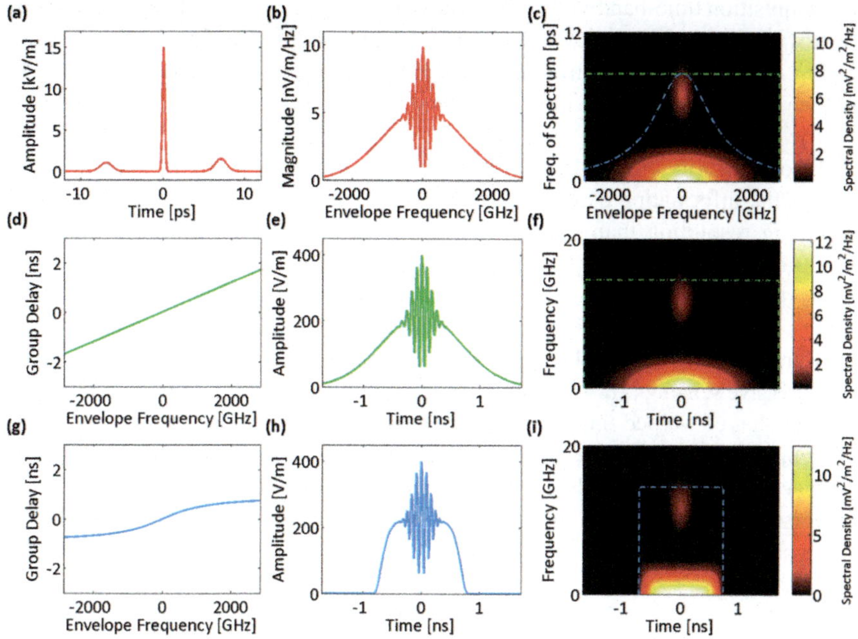

Fig. 10.2 Group delay design based on spectrotemporal sparsity at the spectrum peripheries. (**a**) Envelope of the electric field of an input optical signal. (**b**) The spectrum magnitude of the input envelope. (**c**) Spectrogram of the spectrum magnitude formed by short-term Fourier transform. (**d**) If a temporally dispersive element with a linear group delay profile over the optical bandwidth is used to stretch the input optical field, (**e**) the spectrum maps uniformly to temporal envelope of the electric field. (**f**) Spectrogram of the envelope waveform amplitude resembles that of the spectrum magnitude. (**g**) If a nonlinear group delay profile with lower dispersion at the sides of the bandwidth is used to stretch the optical pulse, (**h**) the spectrum is nonlinearly mapped to the electric field envelope in time. (**i**) The spectrogram of the electric field envelope amplitude after the nonuniform dispersion shows that a shorter temporal window is required to capture the waveform with the same acquisition bandwidth. The *green and blue dot-dash boxes* in Fig. 10.4f, i show the acquisition time and bandwidth, and those in Fig. 10.4c show the effective bandwidth for linear case and nonlinear kernel design, respectively

in Fig. 10.2g. The frequency-to-time mapping and temporal stretching are warped in such a manner that the sparse wings of the spectrum are squeezed relative to the dense central region (Fig. 10.2h). The sparse wings are squeezed in so that they occupy a shorter time duration after the frequency to time mapping. This is desirable as the sides of the spectrum magnitude do not have fast oscillations, and to capture them with a limited acquisition bandwidth, there is no need to stretch them as much as the central part. Essentially, warped stretch avoids overstretching of the spectrum peripheries unlike the linear profile (see Fig. 10.2e). As a result, the reshaped spectrotemporal distribution (Fig. 10.2i) has the same bandwidth as the linear case, but a compressed time duration (compare dot-dashed boxes of Fig. 10.2f, i). We will show that the blue dot-dashed box in Fig. 10.2i, which depicts the acquisition time-bandwidth limit, translates into a frequency-dependent effective bandwidth on the frequency of spectrum as shown with the blue dot-dash contour in Fig. 10.2c, and the spectrotemporal feature sparsity of the signal can be used to reduce the envelope acquisition time-bandwidth product in time stretch dispersive Fourier transform.

Upon uniform temporal sampling, the nonuniform mapping of warped stretch causes the information dense portion of the spectrum to effectively receive higher sampling resolution than the information sparse regions, leading to nonuniform spectral sampling. The local spectral sampling rate is basically designed to match the signal's spectrum sparsity. We note that this nonuniform sampling is performed not by a hard-to-reach variable rate sampler, but with a uniform sampler preceded by warped spectrotemporal reshaping. This approach offers similar functionality as compressive sensing [28, 154] albeit it achieves it via an entirely different approach, namely that of warped time stretch dispersive Fourier transform. If the input signal is significantly chirped, the group delay profile should correspond to the difference between the desired profile for the transform limited version of the input signal (inverse Fourier transform of the input signal spectrum modulus) and the chirp of the input signal.

10.2.1 Spectral Resolution

The resolution of the nonuniform spectral sampling is determined by the sampling resolution of the temporal waveform and the spectral resolution of dispersive Fourier transform. More specifically, the resolution of the spectral sampling using time stretch dispersive Fourier transform is the maximum of resolution limits imposed by the temporal resolution of the photodetector, the bandwidth of the analog-to-digital converter, and the ambiguity in the frequency-to-time mapping of the dispersive Fourier transform. The temporal resolution of the samples is itself limited by the photodetector electrical bandwidth and the Nyquist bandwidth of the analog-to-digital converter. If we assume a resistor-capacitor circuit model for the photodetector output and use its 10–90% rise time as the temporal resolution [163], the photodetector spectral resolution limit is

$$\delta\omega_{\mathrm{PD}}(\omega) = \frac{0.35}{B_{\mathrm{PD}} \cdot \left| \dfrac{\mathrm{d}\tau_g(\omega)}{\mathrm{d}\omega} \right|} \tag{10.1}$$

where $\tau_g(\omega)$ is group delay profile and B_{PD} is the electrical bandwidth of the photodetector. Also, the resolution limit in Fourier domain set by the Nyquist bandwidth of the analog-to-digital converter [164], B_{ADC}, is

$$\delta\omega_{\mathrm{ADC}}(\omega) = \frac{0.5}{B_{\mathrm{ADC}} \cdot \left|\frac{\mathrm{d}\tau_g(\omega)}{\mathrm{d}\omega}\right|} \tag{10.2}$$

The spectral resolution of the dispersive Fourier transform imposed by ambiguity in frequency-to-time mapping is

$$\delta\omega_{\mathrm{DFT}}(\omega) = \sqrt{\frac{4\pi}{\left|\frac{\mathrm{d}\tau_g(\omega)}{\mathrm{d}\omega}\right|}} \tag{10.3}$$

which is derived by stationary phase approximation [47]. Finally, the overall spectral sampling resolution is limited by the largest of these three at each frequency. Therefore,

$$\delta\omega_{\mathrm{Total}}(\omega) = \max\{\delta\omega_{\mathrm{PD}}(\omega), \delta\omega_{\mathrm{ADC}}(\omega), \delta\omega_{\mathrm{DFT}}(\omega)\} \tag{10.4}$$

Clearly, the resolution of the nonuniform spectral sampling is frequency-dependent for a nonlinear group delay profile. Figure 10.3a, c shows the spectral resolutions and their limiting components for both the linear and the nonlinear group delay profiles of Fig. 10.2, respectively (photodetector and analog-to-digital converter Nyquist bandwidths are 14.5 GHz). In low group delay dispersions, temporal resolution limits, $\delta\omega_{\mathrm{PD}}$ or $\delta\omega_{\mathrm{ADC}}$, are mainly the limiting factors, whereas in high group delay dispersions, the resolution of the dispersive Fourier transform dominantly limits the spectral resolution. The spectral sampling resolution can also be translated into an effective bandwidth, B_ω, for the frequency of spectrum. This effective bandwidth, calculated as

$$B_\omega(\omega) = \frac{0.5}{\delta\omega_{\mathrm{Total}}(\omega)} \tag{10.5}$$

is shown with dot-dashed lines in Fig. 10.2c. The effective bandwidth illustrates whether the characteristics of the group delay profile match the spectrotemporal sparsity of the signal, and it proficiently guides the profile design. Figure 10.3b, d shows the group delay profiles of Fig. 10.2 overlaid with their spectral resolutions. The widths of the curves at each frequency correspond to ten times the spectral resolution. One tenth of the overlay width at each group delay corresponds to the set of the optical frequencies that are captured at the same delay and are indistinguishable in the temporal waveform. The magnification factor, ten, is used to make the subtle changes in spectral resolution more noticeable next to the group delay profile, which prominently determines it. The overall spectral sampling resolution of nonlinear group delay profile and the acquisition system is limited by the ambiguity in the frequency-to-time mapping of the dispersive Fourier transform at the center of the spectrum and by the Nyquist bandwidth of the analog-to-digital converter at the peripheries of the spectrum.

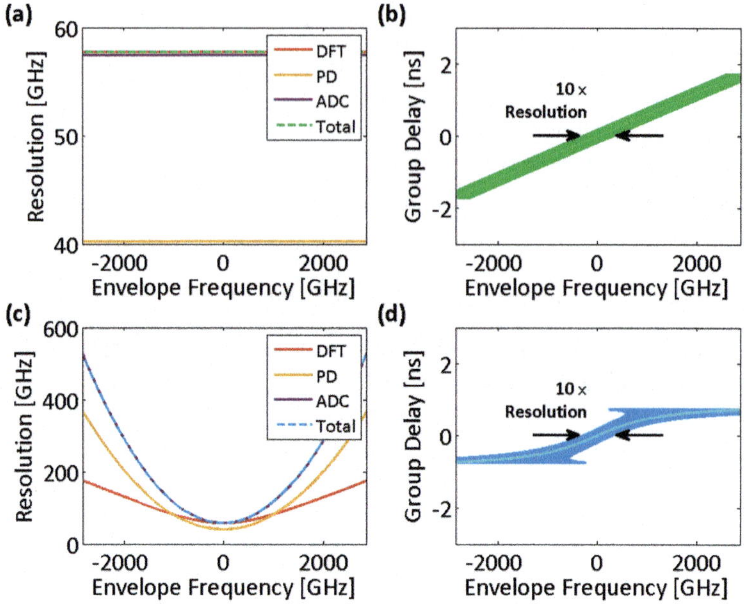

Fig. 10.3 Spectral sampling resolution of time stretch dispersive Fourier transform tuned for the spectrum center. (**a**) Spectral resolution limits for the linear group delay profile and the acquisition system of Fig. 10.2d–f. The overall spectral sampling resolution of the linear time stretch is independent of the envelope optical frequency and limited by the ambiguity in the frequency-to-time mapping of the dispersive Fourier transform. (**b**) The spectral sampling resolution of the linear group delay profile magnified ten times (for visual clarity) and overlapped on the profile. (**c**) Spectral resolution limits for the nonlinear group delay profile and the acquisition system of Fig. 10.2g–i, unlike the linear stretch, depend of the envelope optical frequency. (**d**) Magnified spectral sampling resolution of the warped time stretch overlapped on its group delay profile clearly shows the ambiguity grows at the spectrum peripheries
DFT dispersive Fourier transform, *PD* photodetector, *ADC* analog-to-digital converter, *Total* overall spectral sampling resolution

As another example, we consider a test signal with opposite sparsity compared to that shown in Fig. 10.2. Figure 10.4 shows a signal that is sparse in the central region of the spectrum and feature-dense in the wings. The group delay profile that matches this waveform has higher dispersion, i.e., temporal stretch factor, in the wings. Here, the required acquisition bandwidth is lower for the warped group delay case compared to the linear case. Essentially, the warped profile avoids overstretching of the spectrum center and understretching of the spectrum peripheries, unlike linear profile (see Fig. 10.4e). The spectrogram of the electric field envelope amplitude after the nonuniform dispersion shows that the power at high frequencies is squeezed toward the lower frequencies, and a smaller acquisition bandwidth is required to

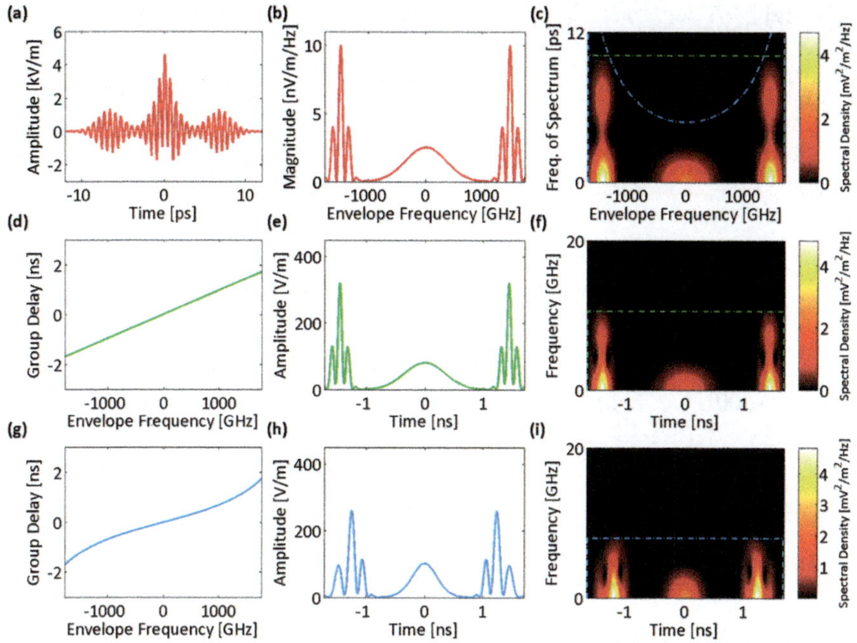

Fig. 10.4 Group delay design based on spectrotemporal sparsity at the spectrum center. (**a**) Envelope of the electric field of an input optical signal. (**b**) The spectrum magnitude of the input envelope. (**c**) Spectrogram of the spectrum magnitude formed by short-term Fourier transform. (**d**) If a temporally dispersive element with a linear group delay profile over the optical bandwidth is used to stretch the input optical field, (**e**) the spectrum maps uniformly to temporal envelope of the electric field. (**f**) Spectrogram of the envelope waveform amplitude resembles that of the spectrum magnitude (shown in Fig. 10.4c). (**g**) If a nonlinear group delay profile with lower dispersion at the center of the bandwidth is used to stretch the optical pulse, (**h**) the spectrum is nonuniformly mapped to the electric field envelope in time. (**i**) The spectrogram of the electric field envelope amplitude after the nonuniform dispersion. The *green and blue dot-dash boxes* in Fig. 10.4f, i show the acquisition time and bandwidth, and those in Fig. 10.4c show the effective bandwidth for linear case and nonlinear kernel design, respectively

capture the waveform with the same temporal sampling duration (blue dot-dash box). This time-bandwidth limit translates into a frequency-dependent effective bandwidth for the frequency of spectrum as shown with the blue dot-dash region in Fig. 10.4c. Figure 10.5 shows the nonuniform spectral resolution corresponding to this group delay design (photodetector and analog-to-digital converter Nyquist bandwidths are 10.5 GHz). The overall spectral sampling resolution of warped stretch transform is limited by the ambiguity in the frequency-to-time mapping of the dispersive Fourier transform at the peripheries of the spectrum and by the Nyquist bandwidth of the analog-to-digital converter at the center of the spectrum.

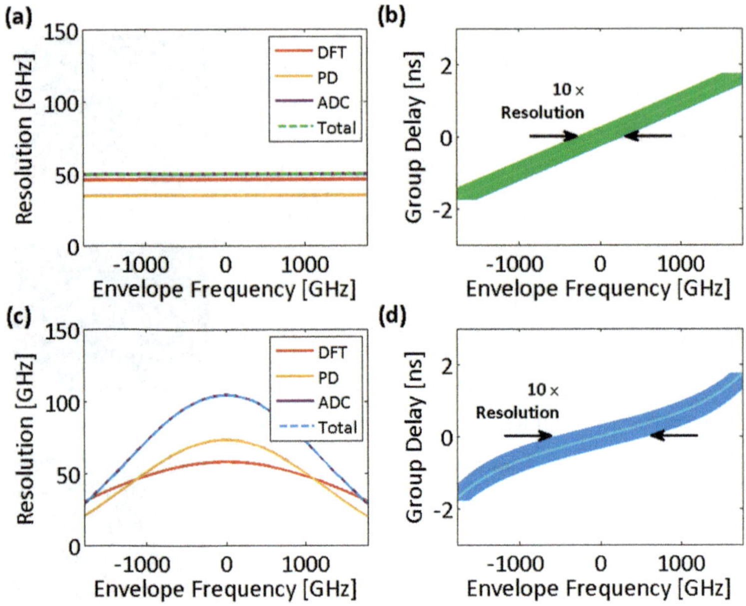

Fig. 10.5 Spectral sampling resolution of time stretch dispersive Fourier transform tuned for the spectrum peripheries. (**a**) Spectral resolution limits for the linear group delay profile and the acquisition system of Fig. 10.4d–f. The overall spectral sampling resolution of the linear time stretch is independent of the envelope optical frequency and limited by the Nyquist bandwidth of the analog-to-digital converter. (**b**) The spectral sampling resolution of the linear group delay profile magnified ten times (for visual clarity) and overlapped on the profile. (**c**) Spectral resolution limits for the nonlinear group delay profile and the acquisition system of Fig. 10.4g–i. (**d**) Magnified spectral sampling resolution of the warped time stretch overlapped on its group delay profile clearly shows the ambiguity grows at the spectrum center
DFT dispersive Fourier transform, *PD* photodetector, *ADC* analog-to-digital converter, *Total* overall spectral sampling resolution

10.2.2 Group Delay Profile Design

The concept of effective bandwidth can be used to design an ideal group delay profile that maximally exploits the spectrotemporal sparsity of a signal. Given an acceptable signal-to-noise ratio (a tolerable spectrotemporal power loss level, e.g., the noise floor of the spectrogram), the important features of the spectrogram can be contoured. At each envelope frequency, the maximum frequency of spectrum on the contour line corresponds to the desired effective bandwidth, B_ω, for the ideal group delay profile. It follows from Eqs. (10.1)–(10.3), and (10.5) that the dispersion of the group delay profile is

$$\left|\frac{\mathrm{d}\tau_g(\omega)}{\mathrm{d}\omega}\right| = \max\left\{\frac{0.7B_\omega}{B_{\mathrm{PD}}}, \frac{B_\omega}{B_{\mathrm{ADC}}}, 16\pi B_\omega^2\right\} \tag{10.6}$$

Since the ideal desired group delay profile is monotonic, it is easily derived as

$$\tau_g(\omega) = \int \left| \frac{\mathrm{d}\tau_g(\omega)}{\mathrm{d}\omega} \right| \mathrm{d}\omega \tag{10.7}$$

The spectrogram that is used here to design the ideal group delay profile does not display the effect of the input signal chirp because it is formed from the spectrum magnitude. If the input chirp is not negligible, it must be subtracted from Eq. (10.7) to get the total group delay profile, which performs the desired frequency-to-time mapping. Figure 10.6a shows a chirped input signal with a spectrum magnitude (Fig. 10.6b) same as that of Fig. 10.4b. The spectrogram is contoured at −30 dB

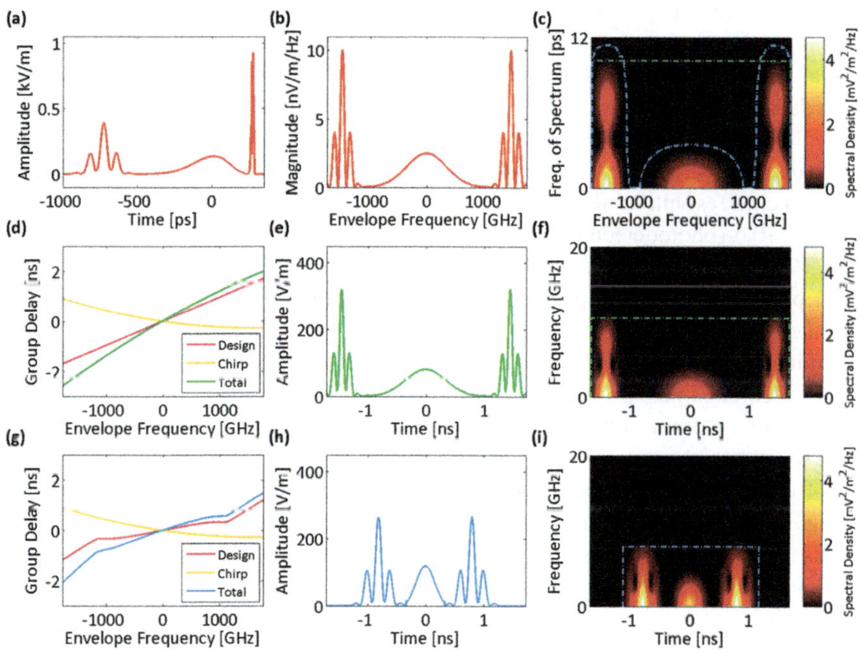

Fig. 10.6 Design of an ideal group delay profile based on spectrotemporal sparsity and input signal chirp. (**a**) Envelope of the electric field of an input optical signal. (**b**) The spectrum magnitude of the input envelope. (**c**) Spectrogram of the spectrum magnitude formed by short-term Fourier transform. (**d**) To perform uniform frequency-to-time mapping, a linear group delay design minus the input signal chirp should be used to stretch the input optical field. (**e**) The spectrum maps uniformly to temporal envelope of the electric field by the total group delay profile. (**f**) Spectrogram of the envelope waveform amplitude resembles that of the spectrum magnitude (shown in Fig. 10.6c). (**g**) If a nonlinear group delay profile is designed based on the *blue dot-dash* contour in Fig. 10.6c and the input signal chirp, (**h**) the spectrum is nonlinearly mapped to the electric field envelope in time. (**i**) The spectrogram of the electric field envelope amplitude after the nonuniform dispersion. The *green and blue dot-dash boxes* in Fig. 10.6f, i show the acquisition time and bandwidth, and those in Fig. 10.6c show the effective bandwidth for linear case and nonlinear kernel design, respectively

of the peak power density (blue dot-dash line in Fig. 10.6c), which specifies the required effective bandwidth of the frequency of spectrum for an ideal group delay profile. We compare using a linear design for the group delay profile (Fig. 10.6d) with designing the ideal group delay profile according to the effective bandwidth by Eq. (10.7) (Fig. 10.6g). The region of the spectrogram with power density 30 dB less than the peak power density is contoured with a blue dot-dash line. This contour is considered as the necessary effective bandwidth to design an ideal group delay profile. In either case, the chirp of the input signal is subtracted from the group delay design to cancel the effect of the input chirp. Thus, the total group delay profiles preserve the desired forms of frequency-to-time mappings (Fig. 10.6e, h). The spectrum is nonlinearly mapped to the electric field envelope in time in such a way that the acquisition time is minimized for the set acquisition bandwidth (8 GHz) and the enforced spectrotemporal accuracy level (−30 dB). Clearly, the power in the sparse regions of the spectrogram is concentrated by the nonlinear profile (Fig. 10.6i) compared to that by the linear profile (Fig. 10.6f), which corresponds to a reduction in the required acquisition time and bandwidth. Essentially, the warped profile ideally avoids overstretching of the spectrum center and understretching of the spectrum peripheries, unlike the linear profile (see Fig. 10.6e). Figure 10.7 shows the spectral resolution of the linear and warped stretch transforms with the profiles in Fig. 10.6 (photodetector and analog-to-digital converter Nyquist bandwidths are 10.5 GHz). The overall spectral sampling resolution of the linear time stretch is independent of the envelope optical frequency and limited by the Nyquist bandwidth of the analog-to-digital converter, while the overall spectral sampling resolution of nonlinear time stretch is limited by the Nyquist bandwidth of the analog-to-digital converter. The linear profile has a uniform resolution across the bandwidth, whereas the resolution of the nonlinear profile is frequency-dependent and designed to match the sparsity of the spectrotemporal distribution.

To further show the applicability of our design method, we consider a signal (Fig. 10.8a) with asymmetric spectrum about the carrier frequency (Fig. 10.8b). This corresponds to a signal with complex temporal envelope (in Fig. 10.8a, we are showing the absolute value of the complex envelope). Using our design algorithm, the spectrogram is contoured at −30 dB of the peak power density (blue dot-dash line in Fig. 10.8c), specifying the required effective bandwidth of the frequency of spectrum for an ideal group delay profile. Note that this contour also becomes asymmetric about the carrier frequency. If we use a chirp-compensated linear group delay profile (Fig. 10.8d), the spectrum to time mapping would be uniform. Of course, this results in an asymmetric temporal waveform (Fig. 10.8e), which resembles the input signal spectrum (Fig. 10.8b). The vertical flip is due to higher frequencies receiving larger group delays, lagging more behind. The frequencies higher than the center frequency experience relatively positive group delays meaning they lag behind the center frequency. The spectrogram of the temporal envelope shows the same type of flip (Fig. 10.8f). If we use the nonlinear group delay profile designed with our technique (Fig. 10.8g) to stretch the signal nonuniformly (Fig. 10.8h), compared to the linear case (Fig. 10.8f), a shorter time duration and a smaller acquisition bandwidth are sufficient (Fig. 10.8i). Clearly, even

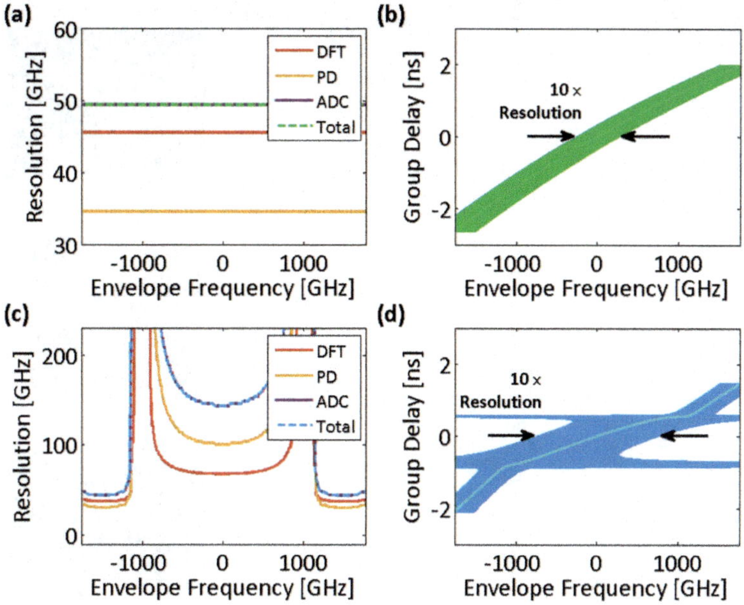

Fig. 10.7 Spectral sampling resolution of time stretch dispersive Fourier transform designed for ideal exploitation of the spectrotemporal sparsity. (**a**) Spectral resolution limits for the chirp compensated linear group delay profile and the acquisition system of Fig. 10.6d–f. (**b**) The spectral sampling resolution of the linear group delay profile minus chirp magnified ten times (for visual clarity) and overlapped on the profile. One tenth of the overlay width at each group delay corresponds to the set of the optical frequencies that are captured at the same delay and are indistinguishable in the temporal waveform. (**c**) Spectral resolution limits for the nonlinear group delay profile and the acquisition system of Fig. 10.6g–i, unlike the linear stretch, depend of the envelope optical frequency. (**d**) Magnified spectral sampling resolution of the warped time stretch overlapped on its group delay profile
DFT dispersive Fourier transform, *PD* photodetector, *ADC* analog-to-digital converter, *Total* overall spectral sampling resolution

in the case of a signal with asymmetric spectrum, the design algorithm leads to a nonlinear group delay profile, which efficiently reduces the time-bandwidth product of the envelope by warped stretch transform. For the group delay profiles shown in Fig. 10.8, the spectral resolutions of the linear and warped stretch transforms are depicted in Fig. 10.9. The overall spectral sampling resolution of the linear time stretch is independent of the envelope optical frequency and limited by the Nyquist bandwidth of the analog-to-digital converter. The linear profile has a fixed resolution across the bandwidth (Fig. 10.9a, b), but the resolution of the nonlinear profile is frequency dependent (Fig. 10.9c, d). For example, the algorithm has designed the nonlinear profile in such a way that it allocates better resolution to the parts of the spectrum that contain fast variations (feature-dense regions).

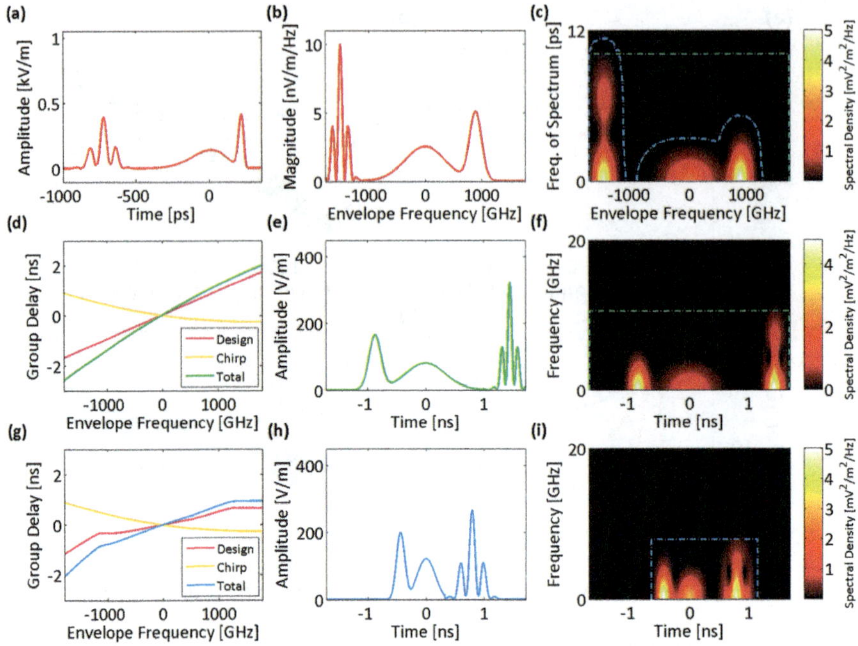

Fig. 10.8 Design of an ideal group delay profile for a signal with asymmetric spectrum about the carrier frequency. (**a**) Envelope of the electric field of an input optical signal. (**b**) The spectrum magnitude of the input envelope is asymmetric. (**c**) Spectrogram of the spectrum magnitude formed by short-term Fourier transform. (**d**) To perform uniform frequency-to-time mapping, a linear group delay design minus the input signal chirp should be used to stretch the input optical field. (**e**) The spectrum maps uniformly to temporal envelope of the electric field by the chirp-corrected linear group delay profile. (**f**) Spectrogram of the envelope waveform amplitude resembles that of the spectrum magnitude (shown in Fig. 10.8c). (**g**) If a nonlinear group delay profile is designed based on the blue dot-dash contour in Fig. 10.8c and the input signal chirp, (**h**) the spectrum is nonuniformly mapped to the electric field envelope in time. (**i**) The spectrogram of the electric field envelope amplitude after the nonuniform dispersion. The *green and blue dot-dash boxes* in Fig. 10.8f, i show the acquisition time and bandwidth, and those in Fig. 10.8c show the effective bandwidth for linear case and nonlinear kernel design, respectively

10.2.3 Simulation Model

We used a discrete-time complex-envelope simulation model for the analysis of warped stretch transform. The carrier frequency is assumed to be 200 THz, resembling an optical wavelength of 1.5 μm. The signal spectra are downshifted from carrier frequency to baseband, easing the required temporal resolution of the simulations. The temporal and spectral resolutions at baseband simulations are set at 0.1 ps and 150.15 MHz, respectively. The output complex envelope spectrum, $\widetilde{E}_{\text{out}}(\omega)$, is calculated as the product of the input complex envelope spectrum, $\widetilde{E}_{\text{in}}(\omega)$, and the impulse response of the downshifted dispersion profile:

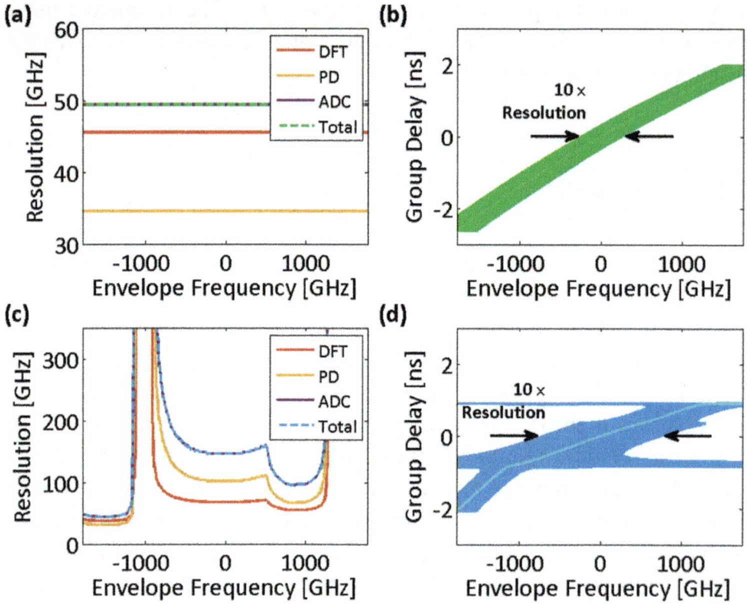

Fig. 10.9 Spectral sampling resolution of time stretch dispersive Fourier transform designed according to the spectrotemporal sparsity for a signal with asymmetric spectral features. (**a**) Spectral resolution limits for the chirp compensated linear group delay profile and the acquisition system of Fig. 10.8d–f. (**b**) The spectral sampling resolution of the linear group delay profile minus chirp magnified ten times (for visual clarity) and overlapped on the profile. (**c**) Spectral resolution limits corresponding to the nonlinear group delay profile and the acquisition system of Fig. 10.8g–i. The overall spectral sampling resolution is limited by the Nyquist bandwidth of the analog-to-digital converter. (**d**) Magnified spectral sampling resolution of the warped time stretch overlapped on its group delay profile clearly shows the ambiguity grows at the regions of the spectrum that do not contain sharp spectral features
DFT dispersive Fourier transform, *PD* photodetector, *ADC* analog-to-digital converter, *Total* overall spectral sampling resolution

$$\widetilde{E}_{\text{out}}(\omega) = H(\omega) \cdot \widetilde{E}_{\text{in}}(\omega) \tag{10.8}$$

Here, ω is the modulation frequency, and the impulse response, $H(\omega)$, is a frequency-dependent phase shift filter;

$$H(\omega) = \exp\{j\varphi(\omega)\} \tag{10.9}$$

where the phase shift, $\varphi(\omega)$, corresponds to the integral of the group delay profile; i.e.,

$$\varphi(\omega) = \int \tau_g(\omega)\mathrm{d}\omega \tag{10.10}$$

Furthermore, linear interpolation is employed to generalize designed group delay to arbitrary frequencies. If any, the chirp of the signal is derived by a moving-window short-time Fourier transform.

10.2.4 Spectrograms

Spectrograms depend on the estimation method of power spectral density, e.g., the window size of the short-term Fourier transform. If the width of the short-time Fourier transform window is reduced, the time resolution of the spectrogram improves, but its frequency resolution degrades. For the frequency-of-spectrum spectrogram, these changes in window size alter the effective bandwidth contour and lead to variations in the design of the group delay profile. In other words, the width of the spectral window should be small enough to capture localized fluctuations of the spectral sparsity, but not too narrow to overestimate the effective bandwidth. Alternatives to spectrogram, for example instantaneous frequency estimated from the analytic form of the signal calculated by the Hilbert transformation, can also be used to determine the required effective bandwidth and design the group delay profile.

10.3 Discussion

The nonuniform sparse Fourier domain sampling described above may be used for data compression. This works when some frequencies carry more information than others. Such frequencies are coded with fine resolution preserving features of spectrum at these frequencies. On the other hand, less important frequencies are coded with a coarser resolution. Naturally, some of the finer details of less important frequencies will be lost in the coding.

The information of interest is usually encoded into the magnitude of the spectrum, therefore a simple unwarping of the time-to-spectrum map is sufficient for reconstruction. For a more general case where the information is contained in both the amplitude and phase, reconstruction requires either coherent detection or recovery of phase from amplitude measurements. The input signal is then recovered by simulation of back propagation through the dispersive profile (filter). Generally known as a phase retrieval method, there are numerous digital algorithms available for recovering the complex amplitude from intensity-only measurements [165, 166].

The reconstruction accuracy and lossy nature of this compression have been analyzed previously [26]. The system reshapes the spectrotemporal structure of the signal such that nearly all the signal energy is within the bandwidth of the photodetector and the real-time digitizer of the acquisition system. Because of the limited resolution of dispersive Fourier transform by ambiguity in frequency-to-time mapping, the limited bandwidths of the photodetector and the digitizer, and the limited resolution of the digitizer, as measured by its effective number of bits (ENOB), the reconstruction will never be ideal, and therefore, this is a lossy compression method. In general, for any time-limited pulse, the spectrum is not bandlimited, and the signal reconstruction will suffer from the loss of out-of-band spectral components in the acquisition system. If the temporal width of the input signal is small enough, so that the spectral resolution is sufficient for capturing details of the input signal spectrum (i.e., in the far field), the bandwidth limitations imposed by the acquisition system can be considered as a frequency-dependent effective bandwidth on the input signal frequency of spectrum as shown in Figs. 10.2c, 10.4c, 10.6c, and 10.8c. The effective bandwidth interpretation facilitates the design of group delay profile for a set of target signals with known spectral characteristics and determines the minimum amount of loss in the compression process.

As a consequence of dispersion, temporal features are transformed and slowed down in time. The amount by which a particular temporal feature is stretched is proportional to the bandwidth of the feature and the overall dispersion over the bandwidth. Fast temporal features have larger bandwidth, and as a result, they are transformed and stretched more than slow temporal features. This feature selective stretch has been referred to as a type of self adaptivity through which the output adapts to the input even when the transfer function of the system is static [22].

The group delay dispersion profile is designed according to the spectral sparsity of the input signal as described, i.e., smart stretching. Beyond that, the group delay is static; in other words, it does not need to be dynamically varied according to the instantaneous behavior of the signal. This is why the time stretch dispersive Fourier transform and its warped counterpart were called self-adaptive [22]. However, if the spectral characteristics of the signal slowly varies over time, a feedback mechanism can be used to adapt the group delay profile, and subsequently the effective bandwidth to the sparsity requirements. The sophisticated group delay profiles designed by our algorithm can be readily implemented by chirped fiber Bragg grating (CFBG) technology [24, 104]. Another implementation option is to use chromo-modal dispersion (CMD) device [103, 160], which uses the large modal dispersion of multimode waveguides in conjunction with the angular dispersion of diffraction gratings to create huge chromatic dispersion. For either technology, the implemented group delay profile will have some deviations from the design. A numerical study of the tolerance to profile nonidealities is performed previously [25].

Time stretch dispersive Fourier transform maps the spectrum of the pulses in a burst-mode signal to the silent intervals in between them. In order to use the time stretch transform for the acquisition of the spectrotemporal evolution of a continuous-time signal, the signal needs to be segmented into multiple pulse trains in a process, which is called virtual time gating [167]. The pulse trains are independently dispersed by linear or warped time stretch systems in parallel, and the acquired signals are digitally stitched together to reveal the spectral features of the continuous-time signal as it varies with time. If the temporal duration of each time gate window is very small, that is, the sliced segments of the continuous-time signal are very short, the resolution of the linear or warped dispersive Fourier transform can be limited by the bandwidth of the gating window. Also, the temporal durations of the time gate windows can be different.

For burst mode signals, time stretch dispersive Fourier transform takes advantage of the empty intervals between pulses to map their spectra into temporal waveforms and captures the signals with a bandwidth far beyond the acquisition bandwidth of the electronic back-end, e.g., analog-to-digital converter and photodetector. Linear time stretch dispersive Fourier transform acquires the spectrum of each pulse with uniform spectral resolution up to frequencies far beyond the electrical acquisition bandwidth of the analog-to-digital converter and photodetector (Fig. 10.10a). Warped time stretch dispersive Fourier transform has the same properties as its linear counterpart, but its spectral resolution is not uniform across the bandwidth (Fig. 10.10b). This nonuniformity can be designed to match the spectrotemporal sparsity of the signal and therefore increase the spectral resolution at desired frequencies under the same envelope time-bandwidth product. Both linear and warped time stretch dispersive Fourier transforms can be used in conjunction with virtual time gating technique for acquisition of the continuous-time signals (Fig. 10.10c, d). In virtually time gated warped stretch transform, gates can have dissimilar group delay profiles corresponding to different distributions of the nonuniform spectral resolution and be suitable for various types of spectral sparsity (Fig. 10.10d). The temporal durations of the gates should not be very short to limit the spectral sampling resolution, but can be different. In virtually time gated time stretch transform, all of the gates (shown with different colors) have the same uniform spectral resolution if identical dispersions and back-end electronics are used.

Spectrotemporal characteristics of a signal can also be analyzed digitally by capturing the signal using an analog-to-digital converter and performing short-time Fourier transform or wavelet transform on the samples. However, in these cases, the maximum frequency that can be measured is limited by the electronic acquisition bandwidth (Fig. 10.10e, f). The wavelet transform can also generate nonuniform temporal resolution for the spectrotemporal distribution of the signal while keeping the spectral resolution fixed (Fig. 10.10f).

10.4 Conclusion

Time stretch dispersive Fourier transform is an indispensable tool for acquisition and analysis of the wideband signals at frequencies far beyond the acquisition bandwidth of the electronic back-end, i.e., digitizer and photodetector. The more general form of it, warped time stretch, offers nonlinear mapping of spectrum to time, which leads to a nonuniform sampling of the spectrum. We analyzed the spectral resolution of the warped time stretch dispersive Fourier transform and defined an effective

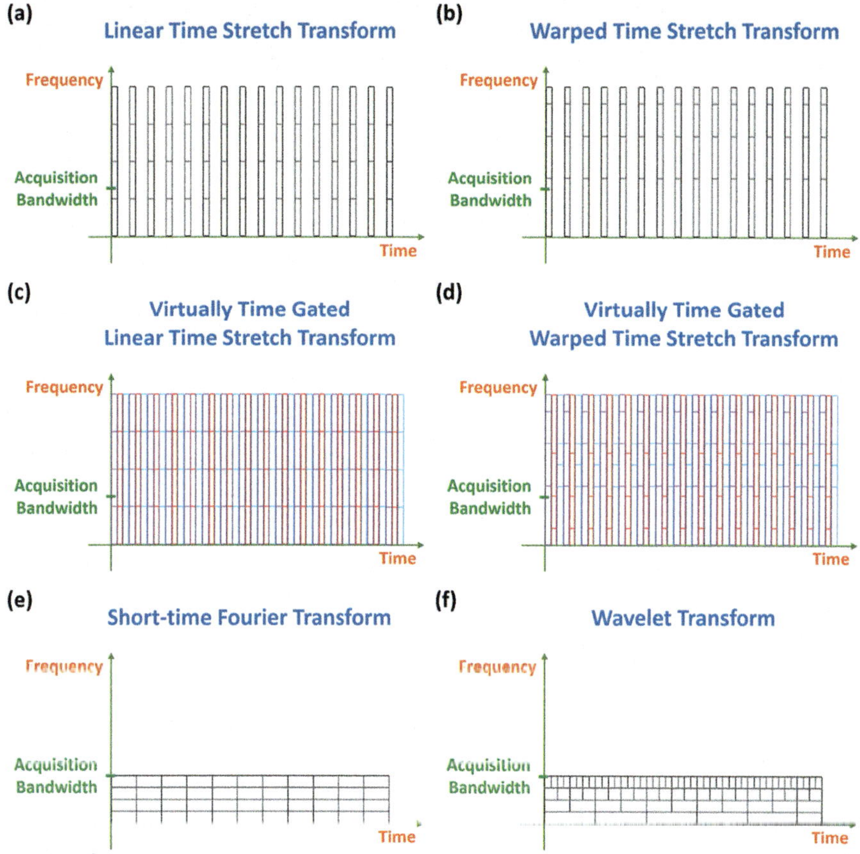

Fig. 10.10 Spectrotemporal resolution of time stretch dispersive Fourier transform vs short-time Fourier transform and wavelet transform. (**a**) In linear time stretch, the spectral resolution is uniform, and the temporal resolution is same as the input pulse width. (**b**) In warped time stretch, the spectral resolution is nonuniform as discussed earlier, and the temporal resolution is again same as the input pulse width. (**c**) For analysis of continuous-time signals, virtual time gating can be used for both linear and warped stretch transforms. (**d**) In addition, for virtually time gated warped time stretch transform, the gates can have different distributions of nonuniform spectral resolutions. (**e**) Short-time Fourier transform can also be used to digitally generate the spectrotemporal distribution of an already acquired signal, but its bandwidth is limited to that of the electronic acquisition system. (**f**) The bandwidth of digitally implemented wavelet transform is also restricted to the electronic acquisition bandwidth, but its temporal resolution can be nonuniform

bandwidth for the transform, which guides the design of a proper group delay profile based on the spectral sparsity of the signal. Finally, linear and warped time stretch transforms are compared to other methods such as wavelet transform for spectrotemporal analysis of continuous-time signals.

Chapter 11
Concluding Remarks and Future Work

In summary, we demonstrated a new type of imaging flow cytometry based on coherent stretched-time-encoded amplified microscopy, namely time stretch quantitative phase imaging (TS-QPI), which is capable of classifying cells with minimal motion distortion at unprecedented rates of 100,000 cells/s. TS-QPI relies on spectral multiplexing to capture simultaneously both phase and intensity quantitative images in a single measurement, generating a wealth of information of each individual cell eliminating the need for labeling with undesirable biomarkers.

To further improve the accuracy and repeatability in label-free phenotypic screening, we introduced a novel machine learning pipeline to label-free cell classification, which harnesses the advantages of multivariate feature detection and deep learning. We demonstrated application of various learning algorithms including deep neural networks, logistic regression, naive Bayes, as well as a new training method based on evolutionary global optimization of receiver operating characteristics (ROC). The results from two experimental demonstrations, one on detection of cancerous cells among white blood cells, and another one on identification of lipid-rich algae, show that classification accuracy by using the TS-QPI hyperdimensional space is more than 17% better than the conventional size-based techniques. Our system paves the way to cellular phenotypic analysis as well as data-driven diagnostics, and thus is a valuable tool for high-throughput label-free cell screening in medical, biotechnological, and research applications.

Furthermore, we demonstrated real-time optical data compression applied to time stretch imaging. This technique provides a solution to the challenges in acquiring and storing the large amount of data generated in high-speed imaging. Using nonlinear group delay dispersion, we achieved warped stretch imaging in such a way that the information-rich central vision is sampled at a higher sample density than the sparse peripheral vision. Most notably, this was done using a uniform electronic sampler, i.e., without adaptive or dynamic control over the electronic sampling rate.

A. Mahjoubfar et al., *Artificial Intelligence in Label-free Microscopy*,
DOI 10.1007/978-3-319-51448-2_11

A three-time image compression was achieved in experimental proof of concept demonstration. Our nonuniform sampling technique could offer one route to taming the capture, storage, and transmission bottlenecks associated with big data.

Finally, we demonstrated how to design the kernel of time stretch transform for acquisition and analysis of the wideband signals at frequencies far beyond the acquisition bandwidth of the electronic back-end, i.e., digitizer and photodetector. We analyzed the spectral resolution of the warped time stretch dispersive Fourier transform and defined an effective bandwidth for the transform, which guides the design of a proper group delay profile based on the spectral sparsity of the signal. In addition, linear and warped time stretch transforms are compared to other methods such as wavelet transform for spectrotemporal analysis of continuous-time signals.

For future work, we recommend implementation of deep learning on cell images to enhance the object recognition and feature extraction in addition to manual feature design. We also suggest integrating information of genomics, proteomics, and phenomics to have better understanding of genotype-to-phenotype mapping. This will enable immediate personalized diagnosis with less complicated tests at phenotypic level.

References

1. Jalali, B., Mahjoubfar, A., & Chen, C. L. (2015). High-throughput biological cell classification featuring real-time optical data compression. *2015 49th Annual Conference on Information Sciences and Systems (CISS)* (p. 7086896). IEEE.
2. Mahjoubfar, A., Chen, C. L., Lin, J., & Jalali, B. (2017). AI-augmented time stretch microscopy. In *SPIE BIOS* (pp. 100760J–100760J). Washington, DC: International Society for Optics and Photonics.
3. Chen, C. L., Mahjoubfar, A., Tai, L.-C., Blaby, I. K., Huang, A., Niazi, K. R., & Jalali, B. (2016). Deep learning in label-free cell classification. *Scientific Reports, 6*, 21471.
4. Mahjoubfar, A., Chen, C. L., & Jalali, B. (2015). Design of warped stretch transform. *Scientific Reports, 5*, 17148.
5. Chan, J. C. K., Mahjoubfar, A., Chen, C. L., & Jalali, B. (2016). Context-aware image compression. *PloS One, 11*(7), e0158201.
6. Chan, J. C. K., Mahjoubfar, A., & Jalali, B. (2016). Optics-inspired context-aware image compression using warped stretch transform. In *2016 IEEE Photonics Society Summer Topical Meeting Series (SUM)* (pp. 214–215). New York: IEEE.
7. Solli, D. R., Ropers, C., Koonath, P., & Jalali, B. (2007). Optical rogue waves. *Nature, 450*(7172), 1054–1057.
8. Ng, W., Rockwood, T., & Reamon, A. (2014). Demonstration of channel-stitched photonic time stretch analog-to-digital converter with enob > 8 for a 10 ghz signal bandwidth. In *GOMACTech* (p. 26.2). Washington, DC: US Department of Defense.
9. Goda, K., & Jalali, B. (2013). Dispersive fourier transformation for fast continuous single-shot measurements. *Nature Photonics, 7*(2), 102–112.
10. Goda, K., Tsia, K. K., & Jalali, B. (2009). Serial time-encoded amplified imaging for real-time observation of fast dynamic phenomena. *Nature, 458*(7242), 1145–1149.
11. Zhang, C., Xu, Y., Wei, X., Tsia, K. K., & Wong, K. K. Y. (2014). Time-stretch microscopy based on time-wavelength sequence reconstruction from wideband incoherent source. *Applied Physics Letters, 105*(4), 041113.
12. Mahjoubfar, A., Goda, K., Ayazi, A., Fard, A., Kim, S. H., & Jalali, B. (2011). High-speed nanometer-resolved imaging vibrometer and velocimeter. *Applied Physics Letters, 98*(10), 101107.
13. Goda, K., Mahjoubfar, A., Wang, C., Fard, A., Adam, J., Gossett, D. R., Ayazi, A., Sollier, E., Malik, O., Chen, E., et al. (2012). Hybrid dispersion laser scanner. *Scientific Reports, 2*, 445.
14. Yazaki, A., Kim, C., Chan, J., Mahjoubfar, A., Goda, K., Watanabe, M., & Jalali, B. (2014). Ultrafast dark-field surface inspection with hybrid-dispersion laser scanning. *Applied Physics Letters, 104*(25), 251106.

A. Mahjoubfar et al., *Artificial Intelligence in Label-free Microscopy*,
DOI 10.1007/978-3-319-51448-2

15. Wei, X., Lau, A. K. S., Xu, Y., Zhang, C., Mussot, A., Kudlinski, A., Tsia, K. K., & Wong, K. K. Y. (2014). Broadband fiber-optical parametric amplification for ultrafast time-stretch imaging at 1.0 μm. *Optics Letters, 39*(20), 5989–5992.
16. Mahjoubfar, A., Goda, K., Wang, C., Fard, A., Adam, J., Gossett, D. R., Ayazi, A., Sollier, E., Malik, O., Chen, E., et al. (2013). 3D ultrafast laser scanner. In *SPIE LASE* (pp. 86110N–86110N). Washington, DC: International Society for Optics and Photonics.
17. Mahjoubfar, A., Chen, C., Niazi, K. R., Rabizadeh, S., & Jalali, B. (2013). Label-free high-throughput cell screening in flow. *Biomedical Optics Express, 4*(9), 1618–1625.
18. Chen, H., Weng, Z., Liang, Y., Lei, C., Xing, F., Chen, M., & Xie, S. (2014). High speed single-pixel imaging via time domain compressive sampling. In *CLEO: Applications and technology* (pp. JTh2A–132). Washington, DC: Optical Society of America.
19. Mahjoubfar, A., Chen, C., Niazi, K. R., Rabizadeh, S., & Jalali, B. (2014). Label-free high-throughput imaging flow cytometry. In *SPIE LASE* (pp. 89720F–89720F). Washington, DC: International Society for Optics and Photonics.
20. Lau, A. K. S., Wong, T. T. W., Ho, K. K. Y., Tang, M. T. H., Chan, A. C. S., Wei, X., Lam, E. Y., Shum, H. C., Wong, K. K. Y., & Tsia, K. K. (2014). Interferometric time-stretch microscopy for ultrafast quantitative cellular and tissue imaging at 1 μm. *Journal of Biomedical Optics, 19*(7), 076001–076001.
21. Diebold, E. D., Buckley, B. W., Gossett, D. R., & Jalali, B. (2013). Digitally synthesized beat frequency multiplexing for sub-millisecond fluorescence microscopy. *Nature Photonics, 7*(10), 806–810.
22. Jalali, B., & Asghari, M. H. (2014). The anamorphic stretch transform: Putting the squeeze on 'big data'. *Optics and Photonics News, 25*(2), 24–31.
23. Asghari, M. H., & Jalali, B. (2013). Anamorphic transformation and its application to time–bandwidth compression. *Applied Optics, 52*(27), 6735–6743.
24. Asghari, M. H., & Jalali, B. (2014). Experimental demonstration of optical real-time data compression. *Applied Physics Letters, 104*(11), 111101.
25. Jalali, B., Chan, J., & Asghari, M. H. (2014). Time–bandwidth engineering. *Optica, 1*(1), 23–31.
26. Chan, J., Mahjoubfar, A., Asghari, M., & Jalali, B. (2014). Reconstruction in time-bandwidth compression systems. *Applied Physics Letters, 105*(22), 221105.
27. Bosworth, B. T., & Foster, M. A. (2014). High-speed flow imaging utilizing spectral-encoding of ultrafast pulses and compressed sensing. In *CLEO: Applications and technology* (pp. ATh4P–3). Washington, DC: Optical Society of America.
28. Valley, G. C., Sefler, G. A., & Shaw, T. J. (2012). Compressive sensing of sparse radio frequency signals using optical mixing. *Optics Letters, 37*(22), 4675–4677.
29. Mahjoubfar, A., Goda, K., Betts, G., & Jalali, B. (2013). Optically amplified detection for biomedical sensing and imaging. *Journal of the Optical Society of America A, 30*(10), 2124–2132.
30. Goda, K., Ayazi, A., Gossett, D. R., Sadasivam, J., Lonappan, C. K., Sollier, E., Fard, A. M., Hur, S. C., Adam, J., Murray, C., et al. (2012). High-throughput single-microparticle imaging flow analyzer. *Proceedings of the National Academy of Sciences, 109*(29), 11630–11635.
31. Kling, J. (2012). Beyond counting tumor cells. *Nature Biotechnology, 30*(7), 578–580.
32. Kærn, M., Elston, T. C., Blake, W. J., & Collins, J. J. (2005). Stochasticity in gene expression: From theories to phenotypes. *Nature Reviews Genetics, 6*(6), 451–464.
33. Carpenter, A. E., Jones, T. R., Lamprecht, M. R., Clarke, C., Kang, I. H., Friman, O., Guertin, D. A., Chang, J. H., Lindquist, R. A., Moffat, J., et al. (2006). Cellprofiler: Image analysis software for identifying and quantifying cell phenotypes. *Genome Biology, 7*(10), R100.
34. Castellini, P., Martarelli, M., & Tomasini, E. P. (2006). Laser doppler vibrometry: Development of advanced solutions answering to technology's needs. *Mechanical Systems and Signal Processing, 20*(6), 1265–1285.
35. Broch, J. T. (1980). *Mechanical vibration and shock measurements*. Copenhagen: Brüel & Kjær.

36. Drain, L. E. (1980). *The laser doppler techniques* (Vol. 1, 250pp.). Chichester, Sussex, New York: Wiley-Interscience.
37. Arnott, W. P., & Sabatier, J. M. (1990). Laser-doppler vibrometer measurements of acoustic to seismic coupling. *Applied Acoustics, 30*(4), 279–291.
38. Goode, R. L., Ball, G., Nishihara, S., & Nakamura, K. (1996). Laser doppler vibrometer (ldv). *Otology & Neurotology, 17*(6), 813–822.
39. Huber, A. M., Schwab, C., Linder, T., Stoeckli, S. J., Ferrazzini, M., Dillier, N., & Fisch, U. (2001). Evaluation of eardrum laser doppler interferometry as a diagnostic tool. *The Laryngoscope, 111*(3), 501–507.
40. Conant, R. (2002). *Micromachined mirrors* (Vol. 12). Berlin: Springer.
41. Pape, D. R., Goutzoulis, A. P., & Kulakov, S. V. (1994). *Design and fabrication of acousto-optic devices*. New York: Dekker.
42. Zheng, W., Kruzelecky, R. V., & Changkakoti, R. (1998). Multichannel laser vibrometer and its applications. In *Third International Conference on Vibration Measurements by Laser Techniques: Advances and Applications* (pp. 376–384). Washington, DC: International Society for Optics and Photonics.
43. Fu, Y., Guo, M., & Phua, P. B. (2010). Spatially encoded multibeam laser doppler vibrometry using a single photodetector. *Optics Letters, 35*(9), 1356–1358.
44. Popescu, G., Ikeda, T., Goda, K., Best-Popescu, C. A., Laposata, M., Manley, S., Dasari, R. R., Badizadegan, K., & Feld, M. S. (2006). Optical measurement of cell membrane tension. *Physical Review Letters, 97*(21), 218101.
45. Goda, K., Tsia, K. K., & Jalali, B. (2008). Amplified dispersive fourier-transform imaging for ultrafast displacement sensing and barcode reading. *Applied Physics Letters, 93*(13), 131109.
46. Qian, F., Song, Q., Tien, E.-K. Kalyoncu, S. K., & Boyraz, O. (2009). Real-time optical imaging and tracking of micron-sized particles. *Optics Communications, 282*(24), 4672–4675.
47. Goda, K., Solli, D. R., Tsia, K. K., & Jalali, B. (2009). Theory of amplified dispersive fourier transformation. *Physical Review A, 80*(4), 043821.
48. Tsia, K. K., Goda, K., Capewell, D., & Jalali, B. (2010). Performance of serial time-encoded amplified microscope. *Optics Express, 18*(10), 10016–10028.
49. Marshall, G. F., & Stutz, G. E. (2011). *Handbook of optical and laser scanning*. Boca Raton, FL: CRC.
50. Fujii, T., & Fukuchi, T. (2005). *Laser remote sensing*. Boca Raton, FL: CRC.
51. Dotson, C., Harlow, R., & Thompson, R. L. (2003). *Fundamentals of dimensional metrology* Albany, NY: Thomson Learning.
52. Göbel, W., Kampa, B. M., & Helmchen, F. (2006). Imaging cellular network dynamics in three dimensions using fast 3d laser scanning. *Nature Methods, 4*(1), 73–79.
53. Pawley, J. (2010). *Handbook of biological confocal microscopy*. Berlin: Springer.
54. Denk, W., Strickler, J. H., Webb, W. W., et al. (1990). Two-photon laser scanning fluorescence microscopy. *Science, 248*(4951), 73–76.
55. Weitkamp, C. (2006). *Lidar: Range-resolved optical remote sensing of the atmosphere*. Springer Science & Business.
56. Schwarz, B. (2010). Mapping the world in 3d. *Nature Photonics, 4*(7), 429–430.
57. Sinha, A. (2010). *Vibration of mechanical systems*. Cambridge: Cambridge University Press.
58. Pelesko, J. A., & Bernstein, D. H. (2002). *Modeling mems and nems*. Boca Raton, FL: CRC.
59. Osten, W. (2006). *Optical inspection of microsystems*. Boca Raton, FL: CRC.
60. Horn, B. (1986). *Robot vision*. Cambridge, MA: MIT.
61. Hoffman, A., Goetz, M., Vieth, M., Galle, P. R., Neurath, M. F., & Kiesslich, R. (2006). Confocal laser endomicroscopy: Technical status and current indications. *Endoscopy, 38*(12), 1275–1283.
62. Tárnok, A., & Gerstner, A. O. H. (2002). Clinical applications of laser scanning cytometry. *Cytometry, 50*(3), 133–143.

63. Vacca, G., Junnarkar, M. R., Goldblatt, N. R., Yee, M. W., Van Slyke, B. M., & Briese, T. C. (2009). Laser rastering flow cytometry: Fast cell counting and identification. In *SPIE BiOS: Biomedical optics* (pp. 71821T–71821T). Washington, DC: International Society for Optics and Photonics.
64. Yaqoob, Z., & Riza, N. A. (2004). Passive optics no-moving-parts barcode scanners. *IEEE Photonics Technology Letters, 16*(3), 954–956.
65. Boudoux, C., Yun, S., Oh, W., White, W., Iftimia, N., Shishkov, M., Bouma, B., & Tearney, G. (2005). Rapid wavelength-swept spectrally encoded confocal microscopy. *Optics Express, 13*(20), 8214–8221.
66. Kelkar, P. V., Coppinger, F., Bhushan, A. S., & Jalali, B. (1999). Time-domain optical sensing. *Electronics Letters, 35*(19), 1661–1662.
67. Chou, J., Boyraz, O., Solli, D., & Jalali, B. (2007). Femtosecond real-time single-shot digitizer. *Applied Physics Letters, 91*(16), 161105–161105.
68. Carlo, D. D. (2009). Inertial microfluidics. *Lab on a Chip, 9*(21), 3038–3046.
69. Grover, W. H., Bryan, A. K., Diez-Silva, M., Suresh, S., Higgins, J. M., & Manalis, S. R. (2011). Measuring single-cell density. *Proceedings of the National Academy of Sciences, 108*(27), 10992–10996.
70. Mrema, J. E., Campbell, G. H., Miranda, R., Jaramillo, A. L., & Rieckmann, K. H. (1979). Concentration and separation of erythrocytes infected with plasmodium falciparum by gradient centrifugation. *Bulletin of the World Health Organization, 57*(1), 133.
71. Chun, J., Zangle, T. A., Kolarova, T., Finn, R. S., Teitell, M. A., & Reed, J. (2012). Rapidly quantifying drug sensitivity of dispersed and clumped breast cancer cells by mass profiling. *Analyst, 137*(23), 5495–5498.
72. Martin, S. J., Bradley, J. G., & Cotter, T. G. (1990). Hl-60 cells induced to differentiate towards neutrophils subsequently die via apoptosis. *Clinical & Experimental Immunology, 79*(3), 448–453.
73. Wyllie, A. H., & Morris, R. G. (1982). Hormone-induced cell death. purification ad properties of thymocytes undergoing apoptosis after glucocorticoid treatment. *The American Journal of Pathology, 109*(1), 78.
74. Wolff, D. A., & Pertoft, H. (1972). Separation of hela cells by colloidal silica density gradient centrifugation i. separation and partial synchrony of mitotic cells. *The Journal of Cell Biology, 55*(3), 579–585.
75. Maric, D., Maric, I., & Barker, J. L. (1998). Buoyant density gradient fractionation and flow cytometric analysis of embryonic rat cortical neurons and progenitor cells. *Methods, 16*(3), 247–259.
76. Bista, R. K., Uttam, S., Wang, P., Staton, K., Choi, S., Bakkenist, C. J., Hartman, D.J., Brand, R. E., & Liu, Y. (2011). Quantification of nanoscale nuclear refractive index changes during the cell cycle. *Journal of Biomedical Optics, 16*(7), 070503–070503.
77. Bosslet, K., Ruffmann, R., Altevogt, P., & Schirrmacher, V. (1981). A rapid method for the isolation of metastasizing tumour cells from internal organs with the help of isopycnic density-gradient centrifugation in percoll. *British Journal of Cancer, 44*(3), 356.
78. Phillips, K. G., Kolatkar, A., Rees, K. J., Rigg, R., Marrinucci, D., Luttgen, M., Bethel, K., Kuhn, P., & McCarty, O. J. T. (2012). Quantification of cellular volume and sub-cellular density fluctuations: Comparison of normal peripheral blood cells and circulating tumor cells identified in a breast cancer patient. *Frontiers in Oncology, 2*, 96.
79. Phillips, K. G., Velasco, C. R., Li, J., Kolatkar, A., Luttgen, M., Bethel, K., Duggan, B., Kuhn, P., & McCarty, O. (2012). Optical quantification of cellular mass, volume, and density of circulating tumor cells identified in an ovarian cancer patient. *Cancer Molecular Targets and Therapeutics, 2*, 72.
80. Gupta, V., Jafferji, I., Garza, M., Melnikova, V., Hasegawa, D. K., Pethig, R., & Davis, D. W. (2012). Apostream^{TM}, a new dielectrophoretic device for antibody independent isolation and recovery of viable cancer cells from blood. *Biomicrofluidics, 6*(2), 024133.

81. Tycko, D. H., Metz, M. H., Epstein, E. A., & Grinbaum, A. (1985). Flow-cytometric light scattering measurement of red blood cell volume and hemoglobin concentration. *Applied Optics, 24*(9), 1355–1365.
82. Liang, X. J., Liu, A. Q., Lim, C. S., Ayi, T. C., & Yap, P. H. (2007). Determining refractive index of single living cell using an integrated microchip. *Sensors and Actuators A: Physical, 133*(2), 349–354.
83. Rappaz, B., Marquet, P., Cuche, E., Emery, Y., Depeursinge, C., & Magistretti, P. (2005). Measurement of the integral refractive index and dynamic cell morphometry of living cells with digital holographic microscopy. *Optics Express, 13*(23), 9361–9373.
84. Curl, C. L., Bellair, C. J., Harris, T., Allman, B. E., Harris, P. J., Stewart, A. G., Roberts, A., Nugent, K. A., & Delbridge, L. (2005). Refractive index measurement in viable cells using quantitative phase-amplitude microscopy and confocal microscopy. *Cytometry Part A, 65*(1), 88–92.
85. Lue, N., Choi, W., Popescu, G., Yaqoob, Z., Badizadegan, K., Dasari, R. R., & Feld, M. S. (2009). Live cell refractometry using Hilbert phase microscopy and confocal reflectance microscopy. *The Journal of Physical Chemistry A, 113*(47), 13327–13330.
86. Gorthi, S. S., & Schonbrun, E. (2012). Phase imaging flow cytometry using a focus-stack collecting microscope. *Optics Letters, 37*(4), 707–709.
87. Vona, G., Sabile, A., Louha, M., Sitruk, V., Romana, S., Schütze, K., Capron, F., Franco, D., Pazzagli, M., Vekemans, M., et al. (2000). Isolation by size of epithelial tumor cells: A new method for the immunomorphological and molecular characterization of circulating tumor cells. *The American Journal of Pathology, 156*(1), 57–63.
88. Di Caprio, G., Schaak, D., & Schonbrun, E. F. (2013). Hyperspectral microscopy of flowing cells. In *Imaging systems and applications* (pp. IM4E–3). Washington, DC: Optical Society of America
89. Barer, R., & Joseph, S. (1954). Refractometry of living cells part i. Basic principles. *Quarterly Journal of Microscopical Science, 3*(32), 399–423.
90. Fard, A. M., Mahjoubfar, A., Goda, K., Gossett, D. R., Di Carlo, D., & Jalali, B. (2011). Nomarski serial time-encoded amplified microscopy for high-speed contrast-enhanced imaging of transparent media. *Biomedical Optics Express, 2*(12), 3387–3392.
91. Boyraz, O., Kim, J., Islam, M. N., Coppinger, F., & Jalali, B. (2000). Broadband, high-brightness 10-gbit/s supercontinuum source for a/d conversion. In *Conference on Lasers and Electro-Optics, 2000. (CLEO 2000)* (pp. 489–490). New York: IEEE.
92. Ikeda, T., Popescu, G., Dasari, R. R., & Feld, M. S. (2005). Hilbert phase microscopy for investigating fast dynamics in transparent systems. *Optics Letters, 30*(10), 1165–1167.
93. Revel, J. P., Hoch, P., & Ho, D. (1974). Adhesion of culture cells to their substratum. *Experimental Cell Research, 84*(1), 207–218.
94. Whur, P., Koppel, K., Urquhart, C. M., & Williams, D. C. (1977). Substrate retention of fractured retraction fibres during detachment of trypsinized bhk21 fibroblasts. *Journal of Cell Science, 24*(1), 265–273.
95. Shapiro, H. M. (2005). *Practical flow cytometry*. New York: Wiley.
96. Watson, J. V. (2004). *Introduction to flow cytometry*. Cambridge: Cambridge University Press.
97. Perfetto, S. P., Chattopadhyay, P. K., & Roederer, M. (2004). Seventeen-colour flow cytometry: Unravelling the immune system. *Nature Reviews Immunology, 4*(8), 648–655.
98. Basiji, D. A., Ortyn, W. E., Liang, L., Venkatachalam, V., & Morrissey, P. (2007). Cellular image analysis and imaging by flow cytometry. *Clinics in Laboratory Medicine, 27*(3), 653–670.
99. Basiji, D. A., & Ortyn, W. E. (2001). Imaging and analyzing parameters of small moving objects such as cells. US Patent 6211955.
100. Razavi, B. (1995). *Principles of data conversion system design* (Vol. 126). New York: IEEE.
101. Zanella, F., Lorens, J. B., & Link, W. (2010). High content screening: Seeing is believing. *Trends in Biotechnology, 28*(5), 237–245.
102. Solli, D. R., Gupta, S., & Jalali, B. (2009). Optical phase recovery in the dispersive fourier transform. *Applied Physics Letters, 95*(23), 231108.

103. Jalali, B., & Mahjoubfar, A. (2015). Tailoring wideband signals with a photonic hardware accelerator. *Proceedings of the IEEE, 103*(7), 1071–1086.
104. Chen, C. L., Mahjoubfar, A., & Jalali, B. (2015). Optical data compression in time stretch imaging. *PLoS One, 10*(4), e0125106.
105. Gires, O., Klein, C. A., & Baeuerle, P. A. (2009). On the abundance of epcam on cancer stem cells. *Nature Reviews Cancer, 9*(2), 143–143.
106. Boddington, S. E., Sutton, E. J., Henning, T. D., Nedopil, A. J., Sennino, B., Kim, A., & Daldrup-Link, H. E. (2011). Labeling human mesenchymal stem cells with fluorescent contrast agents: The biological impact. *Molecular Imaging and Biology, 13*(1), 3–9.
107. Popescu, G. (2011). *Quantitative phase imaging of cells and tissues*. New York: McGraw Hill.
108. Pham, H. V., Bhaduri, B., Tangella, K., Best-Popescu, C., & Popescu, G. (2013). Real time blood testing using quantitative phase imaging. *PloS One, 8*(2), e55676.
109. Feinerman, O., Veiga, J., Dorfman, J. R., Germain, R. N., & Altan-Bonnet, G. (2008). Variability and robustness in t cell activation from regulated heterogeneity in protein levels. *Science, 321*(5892), 1081–1084.
110. Sigal, A., Milo, R., Cohen, A., Geva-Zatorsky, N., Klein, Y., Liron, Y., Rosenfeld, N., Danon, T., Perzov, N., & Alon, U. (2006). Variability and memory of protein levels in human cells. *Nature, 444*(7119), 643–646.
111. Friebel, M., Do, K., Hahn, A., Mu, G., et al. (1999). Optical properties of circulating human blood in the wavelength range 400–2500 nm. *Journal of Biomedical Optics, 4*(1), 36–46.
112. Spencer, S. L., Gaudet, S., Albeck, J. G., Burke, J. M., & Sorger, P. K. (2009). Non-genetic origins of cell-to-cell variability in trail-induced apoptosis. *Nature, 459*(7245), 428–432.
113. Knight, J. B., Vishwanath, A., Brody, J. P., & Austin, R. H. (1998). Hydrodynamic focusing on a silicon chip: Mixing nanoliters in microseconds. *Physical Review Letters, 80*(17), 3863.
114. Lee, G.-B., Chang, C.-C., Huang, S.-B., & Yang, R.-J. (2006). The hydrodynamic focusing effect inside rectangular microchannels. *Journal of Micromechanics and Microengineering, 16*(5), 1024.
115. King, F. W. (2009). *Hilbert transforms* (Vol. 2). Cambridge: Cambridge University Press.
116. Buckley, B. W., Madni, A. M., & Jalali, B. (2013). Coherent time-stretch transformation for real-time capture of wideband signals. *Optics Express, 21*(18), 21618–21627.
117. DeVore, P. T. S., Buckley, B. W., Asghari, M. H., Solli, D. R., & Jalali, B. (2014). Coherent time-stretch transform for near-field spectroscopy. *IEEE Photonics Journal, 6*, 3300107.
118. Driscoll, M. K., Albanese, J. L., Xiong, Z.-M., Mailman, M., Losert, W., & Cao, K. (2012). Automated image analysis of nuclear shape: What can we learn from a prematurely aged cell? *Aging (Albany NY), 4*(2), 119
119. Barer, R., Ross, K. F., & Tkaczyk, S. (1953). Refractometry of living cells. *Nature, 171*(4356), 720.
120. Kamentsky, L., Jones, T. R., Fraser, A., Bray, M.-A., Logan, D. J., Madden, K. L., Ljosa, V., Rueden, C., Eliceiri, K. W., & Carpenter, A. E. (2011). Improved structure, function and compatibility for cellprofiler: Modular high-throughput image analysis software. *Bioinformatics, 27*(8), 1179–1180.
121. Spadinger, I., Poon, S. S. S., & Palcic, B. (1990). Effect of focus on cell detection and recognition by the cell analyzer. *Cytometry, 11*(4), 460–467.
122. Adams, A. A., Okagbare, P. I., Feng, J., Hupert, M. L., Patterson, D., Göttert, J., McCarley, R. L., Nikitopoulos, D., Murphy, M. C., & Soper, S. A. (2008). Highly efficient circulating tumor cell isolation from whole blood and label-free enumeration using polymer-based microfluidics with an integrated conductivity sensor. *Journal of the American Chemical Society, 130*(27), 8633–8641.
123. Nagrath, S., Sequist, L. V., Maheswaran, S., Bell, D. W., Irimia, D., Ulkus, L., Smith, M. R., Kwak, E. L., Digumarthy, S., Muzikansky, A., et al. (2007). Isolation of rare circulating tumour cells in cancer patients by microchip technology. *Nature, 450*(7173), 1235–1239.
124. Gossett, D. R., Weaver, W. M., Mach, A. J., Hur, S. C., Tse, H. T. K., Lee, W., Amini, H., & Di Carlo, D. (2010). Label-free cell separation and sorting in microfluidic systems. *Analytical and Bioanalytical Chemistry, 397*(8), 3249–3267.

125. Maheshri, N., & O'Shea, E. K. (2007). Living with noisy genes: How cells function reliably with inherent variability in gene expression. *Annual Review of Biophysics and Biomolecular Structure, 36*, 413–434.
126. Zangle, T. A., Teitell, M. A., & Reed, J. (2014). Live cell interferometry quantifies dynamics of biomass partitioning during cytokinesis. *PloS One, 9*(12), e115726.
127. Johnston, I. G., Gaal, B., das Neves, R. P., Enver, T., Iborra, F. J., & Jones, N. S. (2012). Mitochondrial variability as a source of extrinsic cellular noise. *PLoS Computational Biology, 8*(3), e1002416.
128. Abu-Mostafa, Y. S., Magdon-Ismail, M., & Lin, H.-T. (2012). *Learning from data*. Seattle: AMLBook.
129. Bishop, C. M., et al. (2006). *Pattern recognition and machine learning* (Vol. 4). New York: Springer.
130. Boddy, L., Morris, C. W., Wilkins, M. F., Tarran, G. A., & Burkill, P. H. (1994). Neural network analysis of flow cytometric data for 40 marine phytoplankton species. *Cytometry, 15*(4), 283–293.
131. Bradley, A. P. (1997). The use of the area under the roc curve in the evaluation of machine learning algorithms. *Pattern Recognition, 30*(7), 1145–1159.
132. Powers, D. M. (2011). Evaluation: From precision, recall and f-measure to roc, informedness, markedness and correlation. *Technical Report*.
133. Huang, J., & Ling, C. X. (2005). Using auc and accuracy in evaluating learning algorithms. *IEEE Transactions on Knowledge and Data Engineering, 17*(3), 299–310.
134. LeCun, Y., Bengio, Y., & Hinton, G. (2015). Deep learning. *Nature, 521*(7553), 436–444.
135. Schmidhuber, J. (2015). Deep learning in neural networks: An overview. *Neural Networks, 61*, 85–117.
136. Hanley, J. A., & McNeil, B. J. (1982). The meaning and use of the area under a receiver operating characteristic (roc) curve. *Radiology, 143*(1), 29–36.
137. Ling, C. X., Huang, J., & Zhang, H. (2003). Auc: A statistically consistent and more discriminating measure than accuracy. In *IJCAI* (Vol. 3, pp. 519–524).
138. Cortes, C., & Mohri, M. (2004). Auc optimization vs. error rate minimization. *Advances in Neural Information Processing Systems, 16*(16), 313–320.
139. Liu, Z., & Tan, M. (2008). Roc-based utility function maximization for feature selection and classification with applications to high-dimensional protease data. *Biometrics, 64*(4), 1155–1161.
140. Verrelst, H., Moreau, Y., Vandewalle, J., & Timmerman, D. (1998). Use of a multi-layer perceptron to predict malignancy in ovarian tumors. *Advances in Neural Information Processing Systems, 10*, 978–984.
141. Merchant, S. S., Kropat, J., Liu, B., Shaw, J., & Warakanont, J. (2012). Tag, you're it! chlamydomonas as a reference organism for understanding algal triacylglycerol accumulation. *Current Opinion in Biotechnology, 23*(3), 352–363.
142. Zabawinski, C., Van Den Koornhuyse, N., D'Hulst, C., Schlichting, R., Giersch, C., Delrue, B., Lacroix, J.-M., Preiss, J.-M., & Ball, S. (2001). Starchless mutants of chlamydomonas reinhardtii lack the small subunit of a heterotetrameric adp-glucose pyrophosphorylase. *Journal of Bacteriology, 183*(3), 1069–1077.
143. Work, V. H., Radakovits, R., Jinkerson, R. E., Meuser, J. E., Elliott, L. G., Vinyard, D. J., Laurens, L. M. L., Dismukes, G. C., & Posewitz, M. C. (2010). Increased lipid accumulation in the chlamydomonas reinhardtii sta7-10 starchless isoamylase mutant and increased carbohydrate synthesis in complemented strains. *Eukaryotic Cell, 9*(8), 1251–1261.
144. Li, Y., Han, D., Hu, G., Dauvillee, D., Sommerfeld, M., Ball, S., & Hu, Q. (2010). Chlamydomonas starchless mutant defective in adp-glucose pyrophosphorylase hyper-accumulates triacylglycerol. *Metabolic Engineering, 12*(4), 387–391.
145. Goodenough, U., Blaby, I., Casero, D., Gallaher, S. D., Goodson, C., Johnson, S., Lee, J.-H., Merchant, S. S., Pellegrini, M., Roth, R., et al. (2014). The path to triacylglyceride obesity in the sta6 strain of chlamydomonas reinhardtii. *Eukaryotic Cell, 13*(5), 591–613.

146. Blaby, I. K., Glaesener, A. G., Mettler, T., Fitz-Gibbon, S. T., Gallaher, S. D., Liu, B., Boyle, N. R., Kropat, J., Stitt, M., Johnson, S., et al. (2013). Systems-level analysis of nitrogen starvation–induced modifications of carbon metabolism in a chlamydomonas reinhardtii starchless mutant. *The Plant Cell Online, 25*(11), 4305–4323.
147. Laudon, M. (2015). Chlamydomonas Resource Center, University of Minnesota. Online.
148. Zhu, Y.-N., Ji, F., Liu, F., Tian, Z.-Q., Zhou, C., & Mahjoubfar, A. (2016). Data mining application in smart meter quality control. In *Fuzzy system and data mining: Proceedings of FSDM 2015* (pp. 369–374). Amsterdam: IOS.
149. Suthar, M., Mahjoubfar, A., Seals, K., Lee, E. W., & Jalali, B. (2016). Diagnostic tool for pneumothorax. In *2016 IEEE Photonics Society Summer Topical Meeting Series (SUM)* (pp. 218–219). New York: IEEE.
150. Zhu, Y., Jian, J., Wu, J., & Yang, L. (2013). Global optimization of non-convex hydro-thermal coordination based on semidefinite programming. *IEEE Transactions on Power Systems, 28*(4), 3720–3728.
151. Chen, C., Mahjoubfar, A., Huang, A., Niazi, K., Rabizadeh, S., & Jalali, B. (2014). Hyper-dimensional analysis for label-free high-throughput imaging flow cytometry. In *CLEO: Applications and technology* (pp. AW3L–2). Washington, DC: Optical Society of America.
152. Wong, T. T., Lau, A. K. S., Ho, K. K. Y., Tang, M. Y. H., Robles, J. D. F., Wei, X., Chan, A. C. S., Tang, A. H., Lam, E. Y., Wong, K. K. Y., et al. (2014). Asymmetric-detection time-stretch optical microscopy (ATOM) for ultrafast high-contrast cellular imaging in flow. *Scientific Reports, 4*, 3656.
153. Nichols, J. M., & Bucholtz, F. (2011). Beating nyquist with light: A compressively sampled photonic link. *Optics Express, 19*(8), 7339–7348.
154. Bosworth, B. T., & Foster, B. T. (2013). High-speed ultrawideband photonically enabled compressed sensing of sparse radio frequency signals. *Optics Letters, 38*(22), 4892–4895.
155. Liang, Y., Chen, M., Chen, H., Lei, C., Li, P., & Xie, S. (2013). Photonic-assisted multi-channel compressive sampling based on effective time delay pattern. *Optics Express, 21*(22), 25700–25707.
156. Chen, Y., Yu, X., Chi, H., Jin, X., Zhang, X., Zheng, S., & Galili, M. (2014). Compressive sensing in a photonic link with optical integration. *Optics Letters, 39*(8), 2222–2224.
157. Wikimedia Commons YY (2007). An artwork by István Orosz.
158. Adam, J., Mahjoubfar, A., Diebold, E. D., Buckley, B. W., & Jalali, B. (2013). Spectrally encoded angular light scattering. *Optics Express, 21*(23), 28960–28967.
159. Di Carlo, D., Irimia, D., Tompkins, R. G., & Toner, M. (2007). Continuous inertial focusing, ordering, and separation of particles in microchannels. *Proceedings of the National Academy of Sciences, 104*(48), 18892–18897.
160. Diebold, E. D., Hon, N. K., Tan, Z., Chou, J., Sienicki, T., Wang, C., & Jalali, B. (2011). Giant tunable optical dispersion using chromo-modal excitation of a multimode waveguide. *Optics Express, 19*(24), 23809–23817.
161. Coppinger, F., Bhushan, A. S., & Jalali, B. (1998). Time magnification of electrical signals using chirped optical pulses. *Electronics Letters, 34*(4), 399–400.
162. Gupta, S., & Jalali, B. (2008). Time-warp correction and calibration in photonic time-stretch analog-to-digital converter. *Optics Letters, 33*(22), 2674–2676.
163. Liu, J.-M. (2005). *Photonic devices*. Cambridge: Cambridge University Press.
164. Oppenheim, A. V., & Schafer, R. W. (2009). *Discrete-time signal processing*. Englewood Cliffs, NJ: Prentice Hall.
165. Walmsley, I. A., & Dorrer, C. (2009). Characterization of ultrashort electromagnetic pulses. *Advances in Optics and Photonics, 1*(2), 308–437.
166. Jaganathan, K., Oymak, S., & Hassibi, B. (2012). Recovery of sparse 1-d signals from the magnitudes of their fourier transform. In *International Symposium on Information Theory Proceedings (ISIT)* (pp. 1473–1477). Cambridge, MA: IEEE.
167. Han, Y., & Jalali, B. (2005). Continuous-time time-stretched analog-to-digital converter array implemented using virtual time gating. *IEEE Transactions on Circuits and Systems I: Regular Papers, 52*(8), 1502–1507.

Index

A. Mahjoubfar et al., *Artificial Intelligence in Label-free Microscopy*,
DOI 10.1007/978-3-319-51448-2